Become a Python Developer

Developer

Wrestle and Defeat It

DOUG PURCELL

Proudly printed in the USA:
ISBN: 0-9973262-9-8
ISBN-13: 978-0-9973262-9-1

Table of Contents

About the author

Hey, this is Doug, aka *Dougie Doug*, the author of this book and founder of Purcell Consult. I want to take the time to thank you for purchasing this book. I would like to stay in contact with you, and the easiest way for me to do this is through my newsletter. I can keep you in the loop about a fresh blog post, an upcoming book, or a cool geek party I'll throw – no guarantees that the last one will happen anytime soon. You can subscribe to my newsletter for the cost of your email address here: http://purcellconsult.com/newsletter

Also, I'm on social media, currently most active on LinkedIn. Go ahead and send me a connect request here: https://www.linkedin.com/in/doug-purcell

Make sure to include a message with your request so that I'll know you're a reader. :-)! My last message to you before you venture off on your road to becoming a highly paid Python developer is to have as much fun as possible. Yeah, programming can be difficult at times but it's also a heck of a lot of fun. If you're having fun while coding then you're doing it right.

ACKNOWLEDGMENTS

FREEDOM, FREEDOM, FREEDOM! The bulk of this document was contrived using free software, so I would like to give a shout-out to all of the software I used and organizations behind it because without it I would have to use something else like Windows. First-off, the operating system I used during the development of the code was **Ubuntu 18.04** (Bionic Beaver). Ubuntu is a Linux distro, developed by Canonical Ltd. You can download it here scot free: https://www.ubuntu.com/download/desktop The word processor that was used to compile the text was LibreOffice which is developed by The Document Foundation. You can download **LibreOffice** here: https://www.libreoffice.org

The diagrams were crafted using **Draw.io** which is an open source technology stack built on top of *mxGraph*, a client side JavaScript diagramming library. Draw.io integrates nicely with Google Drive and Dropbox, and you don't have to download or install any software as the diagramming app is browser based. You can access Draw.io here: https://www.draw.io. I also used Flickr to curate many of the images in this book. It's a swell Yahoo-owned site and you can find yourself plenty of pretty pics to use for a wide spectrum of purposes. Check out Flickr here: https://www.flickr.com

And of course, a shout-out to **Python** the language in which this book is centered around. Without the development of this language many years ago by a clever Dutch programmer named Guido van Rossum, I would sadly had no choice but to choose something else to write about like competitive eating. You can stay up to date with everything happening about Python from their official site: https://www.python.org

The Python Background Check

Python has dramatically risen in popularity in the last several years. Its recent surge may lead most developers to think that it's a *newish* language. However, it's not young by programming standards and the first version of it was released back in 1991, four years earlier than Java which is another popular programming language. Python was the brainchild of Dutch mathematician and computer scientist Guido van Rossum, and was conceived in 1991. According to Wikipedia, *Rossum invented Python because he was bored during Christmas holiday and wanted to create an improved version of the ABC programming language.* Well, that was time well spent. Its name was not influenced by the large carnivorous reptile, but instead by the comedy group *Monty Python*.

Python is a multi platform programming language, therefore you can code with it using a variety of operating systems such as Windows, OS X, and Linux. It's popular due to the fact that it can be used in a variety of ways. Some of the business applications for Python are:

- Web development (Django)
- Scientific computing (SciPy)
- Machine learning(Scikit-learn)
- Neural networks (TensorFlow)
- Browser automation (Selenium)
- Robotics (Raspberry Pi)
- Data mining (Pandas)
- Game development (Pygame)
- Networking (socket module)
- Fintech (fintech)

Python developers are highly desirable on the job market. Seasoned programmers not familiar with it can master it to expedite career opportunities, and aspiring programmers can get acquainted with it to help them land their first job. I will invite you to research *Python jobs* in the search engine of your choice to see the plethora of job opportunities that's available globally for Python developers. Here's some links to shortcut the process:

- LinkedIn Jobs: https://www.linkedin.com/jobs/python-jobsPython Jobs: https://www.pythonjobs.com
- Simply Hired: https://www.simplyhired.com/search
- Remote Python: https://www.remotepython.com

Hopefully the cool things you can do with this language and the copious amount of jobs globally for Python developers will be enough motivation to kick-start your foray into the language.

A link to all of the resources in the book

Links are an important part of the web, and technical books definitely need them to serve as additional guidance for readers. The drawback with links is that years, months, or even days later the resource can disappear which is a liability. One solution is to add the resources to a public repository like GitHub. Download the resources to this book on GitHub in .txt format: http://bit.ly/2PoH5gQ

Also, I've added the resources on my blog which you can access here: http://purcellconsult.com/python-book-resources

This way you'll have access to the freshest links regardless of what format you're reading the book on. BTW, if any of the resources d ecay then do let me know on GitHub or on my blog :-). Muchos gracias.

Chapter I: Python Installation Palooza

Festival of lights Berlin - Laddir Laddir — <u>Photo</u> - <u>CC BY 2.0</u>

The first hurdle to jump over in regards to coding in Python is setting up your environment. The operating system (OS) I'm coding in at the time of publication is Ubuntu 18.04, so if you don't have this then you're effectively out of luck :-). In all seriousness, the three major players in operating systems are Linux, OS X, and Windows. I'll provide installation instructions for all three along with recommended text editors. I'll also provide insights on how to run your first Python program and how to use *pip* which is a popular package manager system.

5

Installing Python on Linux (Ubuntu 18.04)

To see if Python is installed on your machine open up the terminal and type in the following:

```
python
```

You can fire-up the terminal by using the keyboard shortcut: *ctr* + *alt* + *t*. The terminal in Ubuntu 18.04 looks like the following:

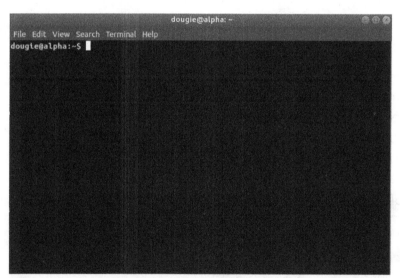

Figure 1.0: Install Python on Linux.

The output should look something like the following:

```
Python 3.6.5 |Anaconda, Inc.| (default, Apr 29
2018, 16:14:56)
[GCC 7.2.0] on linux
Type "help", "copyright", "credits" or "license"
for more information.
```

Look at this line of output:

```
Python 3.6.5 |Anaconda, Inc.| (default, Apr 29
2018, 16:14:56)
```

If you got something like this then *woot-woot*, Python 3.6.5 is installed on your machine. If Python 2.7 or later is installed then it's OK, you don't need to uninstall it, you just need to get Python3 running. Luckily this process is super easy with Ubuntu:

Step one: Open up the terminal by pressing *ctr + alt + t*

Step two: Type sudo apt-get update

Step three: Type sudo apt-get install python3.6

The word *sudo* is abbreviation for "super user do" and it allows programs to be executed as a super user, aka the root user. The *apt* command means *Advanced Package Tool*, which is a package manager for Debian based operating systems like Ubuntu. The a*pt-get command* is the APT package handling utility. You can see a list of the commands that's available for it by typing *apt-get* into the terminal.

Install Python on OS X

Like Linux, Python is already installed on a variety of OS X systems. You can confirm that Python is installed by going to: Applications → Utilities → Terminal. Next, type the following into the terminal:

```
python -V
```

The command will output the version of Python which is:

```
Python 2.7.3
```

Any version between 2.7.0 and 2.7.10 is common. The next step is to test if you have Python3 on your computer. You can do this by typing the following into the terminal:

```
python3
```

If the output shows that *Python 3* is installed then you're safe... for now. If you get an error then that's not cool and you have some work to do. You can fix this by downloading and installing Python with the appropriate Mac OS X installer that matches your system:
https://www.python.org/downloads

Install Python on Windows

It's a high probability that if you're running a Windows operating system then Python won't be there by default. To discover if Python is installed on your machine you can open the terminal and then type *python*. If it's installed then that command will run python.exe and reveal the version number. If you get a rude message like the following:

```
'python' is not recognized as an internal or
external command, operable program or batch file
```

This tells you that Python is not installed and you have to set it up. Follow the steps below to install and setup Python on your computer.

Step one: Download the latest version of Python on your machine: https://www.python.org/downloads

Step two: Open and start the Windows installer that matches your system. If you click "Install Now" then Python is installed in the "user" directory, but if you change its location then make a note of where it's installed.

Step three: You'll have an option to add Python to PATH. In layman terms, the PATH is where the computer searches for Python when you type it via command prompt. If you check this box then Python will be available via this option, if not then when you type *python* in the console an error will occur. Therefore, it's a good idea to check this option so that you can type in python commands via command prompt. If you installed Python without selecting this option then no biggie as you have to manually add the path to your system. Here are the steps on how to add Python to the PATH:

a) In the Windows menu search for advanced system settings and select view advanced system settings.
b) In the window that displays click Environment Variables.
c) In the next window, find and select the user variable called path and click Edit.
d) Scroll to the end of the value and add a semicolon (;) followed by the location of python.exe. If you didn't change the default installation location it should be located in your user directory.
e) Click OK to save the settings.

If you don't know the location of python.exe then don't panic, just search for python.exe in the Windows menu. Once located, right click the file, select properties, and view the Location. Right click to copy the full path and then paste it at the end of the Path user variable. If you don't have a Path user variable then click the *new* button, add a variable named Path, and then add the value which is the location or "path" of the python.exe file. Once done type "python" into the terminal to ensure that everything was set up properly and that it runs.

Hello Peeps!

Once Python is installed we can test a simple Hello Peeps program to whet our appetite with the language. Open up a text editor on your operating system – anything bare bones would do like Notepad on Windows, TextEdit for OS X, or Gedit in Ubuntu. Open up a blank text file and add the following snippet:

```
print("Hello peeps!")
```

Go ahead and save the file as HelloPeeps.py to a location of your choice. It can be anywhere, just don't forget where you put it... pinky swear? The next step is to fire up the terminal or command prompt (if using Windows) and then change into the directory where HelloPeeps.py is located. To change directories use the *cd* command. So, if HelloPeeps.py is in a Programs folder on your Desktop in Ubuntu then it should look something like this:

```
cd Desktop/Programs
```

Once you're in the directory where your Python file is located the next step is to run the program by using the following command:

```
python HelloWorld.py
```

Python is called an **interpreted language** because the programs can be run directly. The file is still compiled; it's just done internally or behind the scenes and is an implementation detail of the language. This is different from Java which if run through the terminal must be explicitly compiled first, and then the byte code is interpreted by the Java Virtual Machine.

You can also run Python code directly through the shell so that it doesn't have to be added to a file and then run – this is convenient when you want quick feedback and feeling too lazy to type code into a text editor. This is officially known as the *Python Shell* and is what we'll be using to learn the fundamental concepts about Python. You can access the Python Shell by opening up the terminal and typing python3. If successful the following will appear:

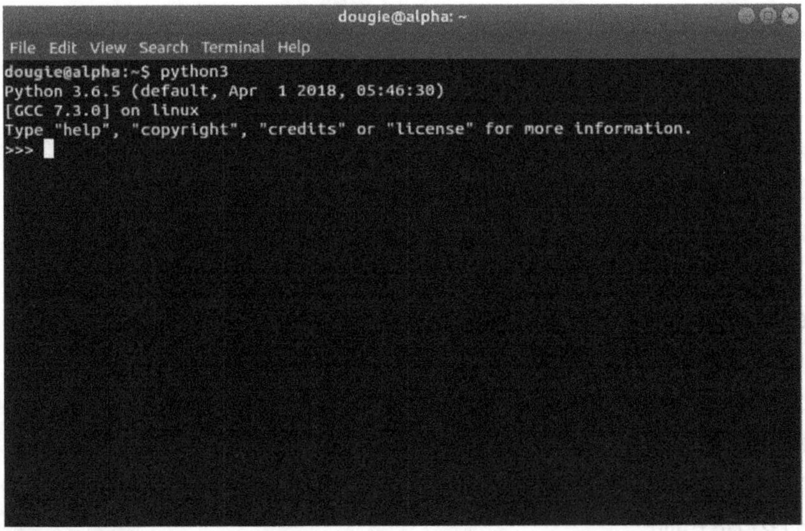

```
dougie@alpha: ~
File  Edit  View  Search  Terminal  Help
dougie@alpha:~$ python3
Python 3.6.5 (default, Apr  1 2018, 05:46:30)
[GCC 7.3.0] on linux
Type "help", "copyright", "credits" or "license" for more information.
>>>
```

Figure 1.1: The Python Shell.

If all is good then simply type the following into the terminal:

```
>>> print("Hello World!")
```

The output will be:

```
Hello World!
```

How to install packages with Pip

Once you got Python installed and tested on your machine

then great, you're one step closer to coding in Python. However, this is not the end to all of your installation drama, in matter of fact it's more like the beginning. In the course of your Python learning experience there will be plenty of resources that you'll need to install that's not included with Python by default. This can be kind of annoying, but the good thing is there's software that helps you install other software. What you'll need is the help of a package manager, and one popular tool for this is *pip*.

Depending on what flavor of Python you're running pip should be preinstalled. More specifically, it comes preinstalled on Python 2.7.9 and later, and Python 3.4 and later (pip3). If you're using an older version of Python I would highly recommend against that as the code crafted in this book is for Python version 3.6+. However, if you insist of using an older version of Python you can follow these steps to installing pip:

1) Download get-pip.py: `https://bootstrap.pypa.io/get-pip.py`

Or, you can use *curl* to download pip by using the following command:

`curl https://bootstrap.pypa.io/get-pip.py -o get-pip.py`

2) Run `get-pip.py`. You can do this by using the following command: `python get-pip.py`

That's it! Below are the steps on how you can install pip on OS X and various Linux environments.

OS X

```
sudo easy_install pip
```

Debian/Linux

```
apt install python3-pip
```

Fedora

```
dnf install python3
```

CentOS

```
yum install python-pip
```

Pip is a command line tool, so to install a package for example you just type a command into the terminal. Here's a quick rundown of some of the key functionality of pip. To install a package use the follow command:

```
$ pip install SomePackage
```

To uninstall a package use the following command:

```
$ pip uninstall SomePackage
```

To upgrade a package do:

```
$ pip install —upgrade SomePackage
```

To see a list of packages that's outdated use the following command:

```
$ pip list —outdated
```

The Python Interpreter, IDLE, and PyCharm

Now that we have Python installed and learned about pip, it's time to make a decision. What will we use to build our code? Part of the problem is that there are so many options to choose from. I mean do we use an editor like Vim, or a full blown IDE like Wing? It can be overwhelming so here's a quick prescription. Test the waters with the Python Interpreter so that you can *whet your appetite* with the language.

This is a short term solution, and one advantage is that you can type in Python code and then get instant feedback from the interpreter. From there move to a lightweight IDE like IDLE. It's bundled with Python out of the box and its minimal features makes it easy to get the hang of. As you gain more experience graduate to a more feature-rich IDE like PyCharm. Below are tutorials to bring you up to speed with the Python Interpreter, IDLE, and PyCharm:

- The Python Interpreter and IDLE Tutorial:
  http://bit.ly/2OKbSU9
- Getting the Hang of PyCharm:
  http://bit.ly/2PZOB6Z

Chapter I Optimized

This chapter shows you how to setup Python on your machine. Python is a cross platform programming language so it can be run on top of Windows, OS X, or Linux environments. In addition to showing how to setup Python you also learned how to use pip on your machine – this is a command tool for managing Python packages as Python contains a larger ecosystem of software packages that can extend the functionality of Python. Many of the packages for Python can be found in the Python Package Index (PyPI), which is the official third-party software repository for Python.

Chapter II: The Monster Crash Course in Python

Monster: Art in Bloom 2009: Storming Party in Art Museum - Seongbin Im - Image - CC BY 2.0

I pretty much condense the core features of Python into a single chapter so that you can quickly get a grasp of the language. After this chapter you should become familiar with the built in data types, common operators, variables, strings, functions, data structures, control flow, iteration, modules, and classes. The point of this chapter is to equip you with the skills needed to start building your own Python programs. Also, I apologize in advance if the PAC-man-like ghost image is too grisly for you ;).

Variables in Python

A **variable** is a placeholder for data that's changeable. Some examples of changeable data in the real world are day of the week, temperature, and your mood. Variables can hold a myriad of data types such as boolean or numeric. Python is a dynamically typed language, and therefore you can declare variables without explicitly stating a type and it will still compile just dandy. The following code snippets are all legal in Python:

```
>>> a = 1
>>> b = 1.27398202
>>> c = "Jambo"
>>> d = '/0024'
>>> e = [ ]
```

Variables in computer programming are similar to variables in mathematics. For example, when creating equations you may have something of the following nature:

$$5x + 5y + 10 = 100$$

In the above equation, x, and y are variables which means that their values are changeable and thus can fluctuate. This is different from statically typed languages like Java which will generate a compile error if the type of variable is not explicitly defined. For example, below would all be illegal variable names in Python while in Java it will be _A-ok_:

```
int a = 1;
double b =  1.27398202;
String c = "Jambo";
char d = d = '/0024';
```

However, there are some rules that Python programmers must follow when creating variable names:

- Variables must be assigned. For example a = 5000 is allowed, but simply writing a without assigning it is illegal.
- The = sign is called the assignment operator, and inserts the value on the right of the equal sign to that of the variable name.
- Variable names in Python may not start with a number. For example, 1c = "Hello" is illegal, while c1 = "Hello" is OK.
- Underscore is allowed in variable names. For example, the following are allowed:

    ```
    >>> mymesaage_ = "hey"
    >>> my_message = "wake up"
    >>> _my_message = "wake up"
    >>> mymessage_ = "wake up"
    ```

- Variable names are case sensitive as a = 5, and A = 10 are allowed.
- It's legal to reassign variables. For example, the following code snippet is allowed and legal:

```
>>> a = 1
>>> a = 1.1
>>> a = 'c'
>>> a = "c"
>>> a = [1, 1.1, 'c', "c"]
>>> a
[1, 1.1, 'c', 'c']
```

To see the list of rules for naming variables in Python check out PEP (Python Enhancement Proposals):
http://bit.ly/2QFcQTd

Hardwired data types and operators

Python has several standard types and operators that are built into its core. The principle built-in types are: numerics, sequences, mappings, classes, instances, and exceptions. This section will provide an overview of these types and operators along with how to effectively use them when coding.

Boolean Operators

Just think of booleans as truth detectors as they uncover the truth of a statement. Any object can be tested for truthfulness and plugged into a conditional statement. Operations and functions which return a boolean result always returns 0 or False, or 1 for True. Below is a list of boolean operations ordered by ascending priority.

Operation Result

a or b if a is false, then b, else a
a and b if a is false, then a, else b
not a if a is false, then True, else False

Note, Python IS a case-sensitive language. False, is NOT the same as false, so it's important to get the syntax correct or else your compiler will mercilessly nag you. The or/and operators are **short-circuit** operators which mean that the second operator is only evaluated under certain circumstances. For example, in *a or b*, if a evaluates to True then *b* is not evaluated because only one truth statement is needed to render the statement True.

This is a syntactic difference between booleans in C-styled languages. In Java for example, you use ampersand (&&) instead of and, double pipes (||) instead of or, and an exclamatory mark (!) instead of not. If you're transitioning from a C-style language to Python then this may seem unnatural at first but you'll get the hang of it with practice.

Comparison Operators

When writing code you'll most likely need to be comparing things to each other, which is where these eight-handy-comparison operators come into play. The following is a summary of the comparison operators in Python:

Operation	Meaning
<	less than
<=	less than or equal to
>	greater than
>=	greater than or equal to
==	equal
!=	not equal
Is	object identity
is not	negated object identity

Figure 2.0 comparison operators

Objects of different types except numeric types never compare to equal. The following code snippet shows how to use some of the above operators in the following conditional:

```
a = 2
b = 5
if a > b and b > 10:
    print("uno")
else:
    print("dos")
...
dos
```

Numeric types

Python has three distinct numeric types which are integers, floating point numbers, and complex numbers. Booleans which was previously discussed are a subtype of integers. **Integers** are negative or positive whole numbers and they have unlimited precision – Integers in Python are similar to integers in mathematics. For example, 2526, -209922, and 829205 are all valid integers in Python. **Floating point numbers** are numerals that can be represented in fractional notation. Examples of floating point numbers in mathematics are real numbers which include integers, fractions like $\frac{4}{3}$, and irrational numbers like the popular π. **Complex numbers** have a real and imaginary part which each is a floating point number. A complex number in mathematics is a number that can be expressed in the form of a + bi, where a and b are real numbers, and i is a solution of the equation $x^2 = -1$.

- You can extract the real and imaginary part of a complex number c, by using c.real and c.imag. When you add 'j' or 'J' to a numeric literal this creates an imaginary number... in this case a complex number with real and imaginary parts. You can use the int(), float(), and complex() constructors to create a number of a specific type. Below is a table of some of the operations that can be applied to numeric types.

Operation	Result	Example
a + b	addition	5 + 6 = 11
a - b	subtraction	20 – 15 = 5
a * b	multiplication	4 * 6 = 24
a / b	division	5 / 2 = 2.5
a // b	floor	6.25 // 2 = 3.0
a % b	modulus	5 % 3 = 2
abs(a)	absolute value	abs(-100) = 100
int(a)	a converted to int	int(2.3) + int(3) = 5
float(a)	a converted to float	float(2) + float(1) = 3.0
complex(re, im)	complex number with real part re, and complex number im	complex(5) = (5 + 0j)
a.conjugate()	conjugate of the complex number of a	5 + 10j.conjugate() = (5-10j)
divmod(a, b)	The pair a // b, a % b	divmod(5,2) = (2, 1)
pow(a, b)	a to power of b	pow(5,3) = 125
a ** b	a to the power of b	5**2 = 25

Figure 2.1: numeric operations in Python.

Built-in data structures

Python has several common data structures such as lists, tuples, sets, dictionaries, and strings. Many of them seem similar like lists and tuples but they have their nuances which will be explored momentarily...

Lists

Python has a built in data type known as a **list** that's a mutable, or changeable ordered sequence of elements. A list could contain numbers, strings, booleans, tuples, sets, or other lists as elements. A list is similar to arrays in that they store data and can access elements using subscript notation ([]), but they also have more specialized properties that make them suitable for scientific computing. To create a list use the square bracket [] notation:

```
[]
[2, 4, 6, 8, 10]
[True, False, True, True, False]
["hip hop", "rock and roll", "country", "electronic"]
```

To access an element in an array you can use the subscript notation, or:

```
list[index]
```

The first element corresponds to the index of 0. Let's take the following list as an example:

```
>>> list = [12, 382, 29, True, False, "Hello", 23.292]
>>> list[2] * 2
```

58
```
>>> list[4] = True
>>> list
[12, 382, 29, True, True, 'Hello', 23.292]
```

Also, list elements can be accessed using negative list notation. The notation `list[-1]` will allow you to access the last element.

```
>>> list[-1]
23.292
```

In this example the length of the list is 7, so `list[-7]` will get you the first element.

```
>>> list[-7]
12
```

It's important to keep track of the accessible indexes in the list. For example, if you try to access an index that doesn't exist an error will occur, like `list[7]`.

```
>>> list[7]
Traceback (most recent call last):
File "<stdin>", line 1, in <module>
IndexError: list index out of range
```

The length of a list is the total amount of elements that it contains, and you can access it by using the `len()` function. Therefore, to get the length of a list, use `len(characters)` which is equal to 5. To get the last element of the list you can do:

```
>>> characters[len(characters) - 1]
```

Here's a diagram that visually represents the following list in

Python:

```
>>> numbers = [29101, 7.229, True, "Hi", 'A']
```

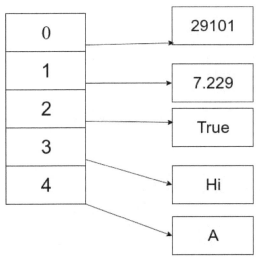

Figure 2.2: Python List Visual.

They say a picture is worth a thousand words so ideally the diagram clarifies lists for you. The list in Python is considered a **sequence type**, and sequence types in Python have a set number of operations. Some of these operations are the plus (+) operator which leads to concatenation and (*) which is for repetition. Below is a list of some of the operators that can be applied to sequence types.

```
>>> x = [2, 622, 8768, 981, 90]
>>> 622 in x
True
>>> 100 not in x
True
```

Lists can be concatenated or combined with the plus (+) operator.

```
>>> a = [1, 2, 3]
>>> b = [4, 5, 6]
>>> a + b
[1, 2, 3, 4, 5, 6]
```

Lists can be multiplied by using the asterisk (*) operator.

```
>>> a = [1, 2]
>>> b = a * 3
>>> b
[1, 2, 1, 2, 1, 2]
```

You can get the length of a list by using the len() method and you can also get the smallest and largest item of a list by using the min() and max() methods appropriately.

```
>>> a = [2, 4, 6, 8, 10]
>>> len(a)
5
>>> min(a)
2
>>> max(a)
10
```

To determine the index in which an element appear you can use the index() method.

```
>>> a.index(4)
1
```

Lastly, you can get the total number of occurrences of

an element in a list by using the `list.count()` method as shown below:

```
>>> list = [1, 1, 1, 1, 2, 5, 87, 100]
>>> list.count(1)
4
```

Tuples

A **tuple** is similar to a list as it can group multiple objects together but the main distinction is that it's immutable. This means that an element can't be added to an existing tuple, but another variable is created that point to the same object. Like lists, tuples have built in operators of in, +, and *, and they also have built-in functions like len(), max(), and min(). Similar to lists, tuples can contain homogeneous or heterogeneous data. The elements of a tuple can be accessed by subscript notation and slicing is also permitted. An example of how to create a tuple along with some of their functions is listed below:

```
>>> tuple = (100, 200, 300, 150)
>>> tuple
(100, 200, 300, 150)
>>> tuple[0]
100
>>> tuple[1:3]
(200, 300)
>>> color = ("red", "green", "blue")
>>> color[0] = "gray"

Traceback (most recent call last):
File "<stdin>", line 1, in <module>
TypeError: 'tuple' object does not support item
```

assignment

Oops! Remember, tuples are immutable so once created the elements can't be changed. Tuples like lists can be nested, so the following is OK:

```
>>> a = (1,2,3)
>>> b = (2,1,5)
>>> c = a, b
>>> c
((1, 2, 3), (2, 1, 5))
```

Therefore, if you're writing code in which an immutable object is required then a tuple is a good choice to consider.

Sets

Humans compile sets of various types. For example, a child may collect trading cards and compile all of their unique cards into one set. A car collector may have a set of classic cars, and a movie enthusiast may have a set of *film noir* DVDs. A **set** in mathematics is a collection of distinct objects. For example, the numbers 3, 6, and 9 are a set of numbers and can be represented mathematically with the following notation:

```
{3, 6, 9}
```

Sets are a popular topic in mathematics and are taught to children at a young age. Some topics such as Venn diagrams are fun and easy to get a hang of. In Python you'll want to use sets to represent objects in which uniqueness is prominent. If you have a duplicate object in a set then it will be ignored. There are two ways in which you can create a set in Python:

- set()
- Or, by assigning a variable to {}

Here are some examples of sets in action in Python:

```
>>> a = set()        # empty set
>>> a = {1, 1, 2, 2, 3}
>>> a
{1, 2, 3}
>>> a = set({1, 2, 2})
>>> a
{1, 2}
>>> a = {6, 292, 9820, 5000, 292}
>>> a
{5000, 292, 6, 9820}
```

Some of the operators that you can apply to sets in Python are: subtraction (-), union (|), and (&), and symmetric difference (^) which returns elements from either set but not both. An example of how to use the operators is listed below:

```
>>> a = {1, 1, 2, 2, 0, 56, 98, 6, 5, 5, 77}
>>> b = {1, 2, 3, 4, 5}

>>> a | b
{0, 1, 2, 98, 3, 5, 6, 4, 77, 56}

>>> a & b
{1, 2, 5}

>>> a - b
{0, 98, 6, 77, 56}
```

```
>>> a ^ b
{0, 98, 3, 4, 6, 56, 77}
```

Dictionaries

Another helpful data type in Python is the dictionary. Unlike sequences which are indexed by a range of numbers, dictionaries are indexed by keys which can be any immutable object. Dictionaries in Python are similar to associative memories or arrays in other languages. Therefore, strings, numbers, and even tuples can be keys for dictionaries – lists can't be used I'm afraid since they're mutuable. According to the official Python documentation its best to think of **dictionaries** as *sets of key value pairs with the requirement that keys are unique within one dictionary.*

A pair of curly braces {} creates an empty dictionary, and placing a comma separated list of key/value pairs inside the braces adds initial key/value pairs to the dictionary.

You'll want to use a dictionary when you need to store a value with a key and extract the value given a key. You can delete a key/value pair by using del, and you can return the list of items in a dictionary named d, by performing list(d).

This will return a list of items in insertion order but in order to sort the items you'll need to use the sorted method instead which looks like this:

```
>>> sorted(d)
```

Also, to check if a key is in a dictionary you can use the in keyword. Below is a demo of the dictionary data structure in Python:

```
>>> cars = {'Koenigsegg' : 4800000, 'Ferrari' :
3000000, 'Lamborghini' : 4500000}
>>> cars['Ferrari']
3000000
>>> del cars['Koenigsegg']
>>> cars
{'Ferrari': 3000000, 'Lamborghini': 4500000}
>>> sorted(cars)
['Ferrari', 'Lamborghini']
>>> 'Lamborghini' in cars
True
```

Lists as Stacks

A **stack** is a Last in First Out (*LIFO*) data structure. Just imagine that you're building a pile of your favorite books, and that you're stacking one on top of each other. The very last book you added to the stack will be the first one you take off, as removing the bottom book would lead to a book avalanche. A similar concept is when you're in a cafeteria and there is a stack of trays. The tray that you'll use to place all of your yummy food on will be the most recent tray added to the stack. The same idea applies to Python when dealing with stacks. Lists have a pop() function which allows you to remove the element at the top of the list as indicated in the code snippet below:

```
>>> brand = ["Jiffy pop", "Pop Secret",
"Butterkist", "ACT II"]
```

The above stack can be illustrated as the following diagram:

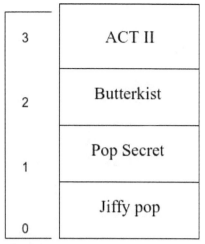

3	ACT II
2	Butterkist
1	Pop Secret
0	Jiffy pop

Figure 2.3 Python List as a Stack.

Below is the result if the following operations were run on it:

```
>>> brand.pop()
'ACT II'
>>> brand.append("pop corn")
>>> brand
['Jiffy pop', 'Pop Secret', 'Butterkist', 'pop corn']
>>> brand.pop()
'pop corn'
>>> brand.pop()
'Butterkist'
>>> brand
['Jiffy pop', 'Pop Secret']
```

List as Queues

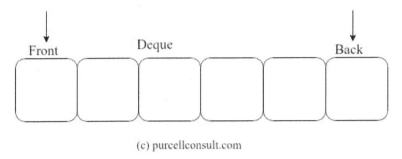

Figure 2.4: Deque illustration.

Queues are similar to stacks except the order is reversed. A queue is like that long line you wait in to eat at the most trendiest place in town. The **queue** is pretty much first come first served, or in other words, First In, First Out (FIFO) order. Now, here's an issue that we need to address. Normal lists are not very efficient for queues. The reason being is that while adding or removing items from the end of the queue is a piece of cake, doing inserts or pops from the beginning is slow or *expensive* because all of the elements must be shifted by one. To implement a queue in Python use collections.deque which was designed for the purpose of fast appending and popping from both ends. Below is an example of how to implement a queue with Lists in Python.

```
>>> from collections import deque
>>> queue = deque(["Charles", "Leo", "Vladimir",
"Ernest", "Virginia"])
>>> queue.pop()
'Virginia'
>>> queue.popleft()
'Charles'
>>> queue.append("George")
>>> queue.insert(1, "Hello")
>>> print(queue)
```

```
deque(['Leo', 'Hello', 'Vladimir', 'Ernest', 'George'])
```

Slicing

Slicing allows you to partition a list within a delineated range. The syntax for slicing a list is shown below:

```
a[i:j] = b
```

Below are examples of how to use slicing on a list.

```
>>> characters = ['a', 'c', 'z', 'e', 'b']
>>> characters[0:3]
['a', 'c', 'z']
>>> characters[0:-2]
['a', 'c', 'z']
>>> food = ["apples", "pasta", "potatoes", "bread",
"milk", "beef"]
>>> food[0:2]
```

Slicing allows you to partition a list within a delineated range. The syntax for slicing a list is shown below:

```
a[i:j] = b
```

Below are examples of how to use slicing on a list.

```
>>> characters = ['a', 'c', 'z', 'e', 'b']
>>> characters[0:3]
['a', 'c', 'z']
>>> characters[0:-2]
['a', 'c', 'z']
>>> food = ["apples", "pasta", "potatoes", "bread",
"milk", "beef"]
>>> food[0:2]
['apples', 'pasta']
```

Here's something to keep in mind. When you slice a list, i is the starting point and j is the endpoint. In a list i is inclusive meaning it's included in the new list, and j, the endpoint is excluded so in this case apples and pasta are the added elements in the new list. Also, be mindful of how to access the ranges of lists as not understanding it can make your program behave strangely. For example, if I wanted to slice food so that the new list contains apples, pasta, and potatoes the code would look like the following:

```
food[0:3]
```

However, let's say that I want to create a new list which contains all of the items of food. Do you think something like food[0:8] would work? The answer is yes, it would print out all of the elements and return the following even though there are only *six* elements in the list:

```
['apples', 'pasta', 'potatoes', 'bread', 'milk', 'beef']
```

Also, like with normal indexes you can also access negative indexes, for example, food[-1] would equal beef. You can also generate subsets of lists by slicing it using a myriad of techniques:

```
>>> food[1:]
['pasta', 'potatoes', 'bread', 'milk', 'beef']
>>> food[:4]
['apples', 'pasta', 'potatoes', 'bread']
>>> food[:]
['apples', 'pasta', 'potatoes', 'bread', 'milk',
'beef']
>>> food[1:-1]
['pasta', 'potatoes', 'bread', 'milk']
```

If you add an additional colon then you can create a subset of the list by defining the step that you can take with each increment.

```
['apples', 'pasta', 'potatoes', 'bread', 'milk',
'beef']
```

Also, like with normal indexes you can also access negative indexes, for example, food[-1] would equal *beef*. You can also generate subsets of lists by slicing it using a myriad of techniques:

```
>>> food[1:]
['pasta', 'potatoes', 'bread', 'milk', 'beef']
>>> food[:4]
['apples', 'pasta', 'potatoes', 'bread']
>>> food[:]
['apples', 'pasta', 'potatoes', 'bread', 'milk',
'beef']
>>> food[1:-1]
['pasta', 'potatoes', 'bread', 'milk']
```

If you add an additional colon then you can create a subset of the list by defining the step that you can take with each increment.

```
>>> b[0:6:2]
[1, 1, 1]
```

Strings and Common Operations in Python

Strings are an important component in virtually every programming language and Python includes a plethora of features for manipulating and processing them. If you're new to programming and never heard of *strings* then just think of it as *a bunch of text*. There are three ways in which you can represent text in Python. You can use single quote strings, double quotes, or triple quotes. For example, the following are all valid ways in which you can create strings in Python.

```
>>> a = "Hello World"
>>> b = """Hello World"""
>>> c = 'Hello World'
>>> d = '''Hello World'''
```

To get a good idea of how strings work let's take a look at the following diagram:

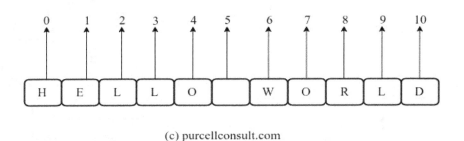

(c) purcellconsult.com

Figure 2.5 The String Index Diagram in Python.

As you can see from the above diagram, a string is like a list of characters. Each character in a string can be accessed at its index which **starts at 0** and goes up to the **length of the string minus one**. Therefore, to access the first element of the string you can do this:

```
>>> a[0]
'H'
```

To access the last element of a string you can use the following notation:

```
>>> a[len(a)-1]
'd'
```

Strings are immutable which means that once they're created they can't be changed. Therefore, once you apply operators on strings such as +, *, and in, then the strings that are returned are actually new strings; in other words it's something of an illusion because the original string is not modified as it's still in memory, but a new string is created and the variable that held the original string now reference a new one. Below is an example of how operators can be applied to strings in Python:

```
>>> "The" + " Cat"
'The Cat'
>>> "Salut, mes amis " * 2
'Salut, mes amis Salut, mes amis'
>>> "i" in "Ciao"
True
>>> "a" not in "Jambo"
False
```

Like lists, strings can also be sliced:

```
>>> message = "Hello World"
>>> message[0:5]
'Hello'
```

Also, strings can be accessed using the negative index. For example, message[-1] returns 'd'.

Mixing Strings with other types

It's important to know how to mix several different types in a string. For example, what if you had to print a string like the following:

```
"Today is the 1st and I have 5 bills due!"
```

It's true that you can simply input all of the numbers in a string and everything will show up as planned like below:

```
'Today is the 1st and I have 5 bills due!'
```

However, the issue with this is that in programming many values are variable and subject to change. For example, this would not fly by if you need to read in user input and then process it into the string.

```
>>> number = input()
print("My lucky number is number")
```

Doing this will lead to the exact input which is:

My lucky number is number

However, what you want is for the actual number to show up so that it'll say:

```
"My lucky number is 10"
```

How exactly do you do this? Well, one way is to **use the comma delimiter** to separate the integer from the text as listed below:

```
>>> number = 10
>>> print("My lucky number is", number)
My lucky number is 10
```

This gets the job done, but if there were more types that

needs to get formatted in the string then this would eventually lead to a long and awkward print statement. Therefore, there are several schools to thoughts when it comes to formatting strings in Python. One, Python borrows a lot from C's `printf()` function. In C, there are a couple of symbols that are used to represent various types in strings. Some of the common symbols are listed in the table below:

- d – placeholder for decimal values
- f – floating point decimal
- c – single character
- s – converts the python object using `str()`
- % – Marks the beginning of the specifier

Let's look at some examples:

```
>>> "Hello %s today is the %dst and I have %d bills
due today" % ("World", 1, 5)

'Hello World today is the 1st and I have 5 bills
due today'
```

There are a lot of interesting things you can do especially when you need to specify the number of digits in an operation. For example, let's assume you have a single digit number like 5. You can specify that it contains 3 digits by doing something like:

```
>>> print("%.3d" % 5)
005
```

If you put the .3 after the d then you'll get 5.3 instead. If you put just the number 3 excluding the dot then you'll get the following result instead:

```
>>> print("%3d" % 5)
```

5

This pads spaces to the left of the number, and if you put the 3 on the right side then it will simply add the number 3 to the right of 5 which equals 53. This can be very helpful when working with floating points. For example, let's assume that you only wanted the first 5 decimals of the following number:

7.29273728722728272828

To do that you'll use the following print statement:

```
>>> print("%.5f" % 7.29273728722728272828)
7.29274
```

If the above is not your cup of tea then you can use an alternative syntax to format strings which makes uses of curly braces:

```
>>> print("{} and {}" .format(5, 5.3839))
5 and 5.3839
```

You can also change the order in which data is printed:

```
>>> print("{2} and {1} and {0}".format(0, 1, 2))
2 and 1 and 0
```

The above may be confusing at first, but just remember that the arguments in the string are placeholders, and the first argument starts at 0, the second 1, and so on. Therefore, in the above code snippet the argument 0 starts at 0, 1 starts at 1, and 2 starts at 2. So, when the curly braces are added with the numbers {2} corresponds to 2 and will display that in the statement. Another way you can format Strings in

Python is with the `format()` function:

```python
def list_factors(n):
    """lists factors of a number n."""
    for x in range(1, n + 1):
        for y in range(1, n + 1):
            if x * y == n:
                print("{0:2d} x {1:2d} = {2:2d}".format(x, y, n))
>> > list_factors(12)

1 x 12 = 12
2 x 6 =  12
3 x 4 =  12
4 x 3 =  12
6 x 2 =  12
12 x 1 = 12
```

The function `list_factors()` prints the factors of a number n in a x b format. The string output can be padded by placing an integer in front of 2. You can play around with the parameters to see how it affects the output of the string.

Raw strings and f-strings

A **raw string** literal is a string literal which is marked by an r or R before its opening quote. The quickest way to see the difference between a traditional string and a raw one is to look at an example:

```python
>>> print('Hello\nWorld')
Hello
World
```

```
>>> print(r'Hello\nWorld')
Hello\nWorld
```

The only difference between these two prints statements is that one contains an r appended to the beginning of it. The \n character is known as an escape character, and these types of characters have alternative meanings; in this case it means to create a newline. However, there are other characters that you could use such as \t for tab, \b for backspace, and \\ for backslash. The one without the r processes the newline and therefore the text are printed on separate lines, while the one with the r appended to the beginning includes the \n character. This is useful if you want the character to appear the way it is in the string. For example, if you wanted to wrap the salutation around quotes using escape characters then one solution is to do the following:

```
>>> print(r'"Hello  World"')
"Hello  World"
```

If you want double quotes then you can simply enclose the string within a single quote and then use double quotes to place around the portion of text you want them to show up at. The f-string is known as **formatted string literals**. This type of string literal has 'f' or 'F' appended to the beginning of it. The main difference between this and the raw string is that formatted strings can do more than just format constants, they can also contain replacement fields which are expressions separated by curly braces. Other strings like raw strings substitute constant value at run time, while formatted string literals evaluate expressions at run time. Here are some examples of *f-strings* in action:

```
>>> name = "Dougie P"
```

```
>>> version = 3
>>> print(f"My name is {name} and I'm here to teach
you Python {version}")

My name is Dougie P and I'm here to teach you
Python 3
```

As shown in the above example {}, curly braces serve as placeholders for strings.

```
>>> text_1 = "is a number"
>>> num_1 = 277272
>>> text_2 = "is another one"
>>> num_2 = 28.292892
>>> print(f'{num_1} {text_1}, and {num_2} {text_2}.')

277272 is a number, and 28.292892 is another one.
```

Difference between Python2 and Python3

A big difference between Python2 and 3 is the print statement as shown below:

Python 2 print statement:

```
>>> print "What is up world!?"
What is up world!?
```

Python 3 print function:

```
>>> print("What is up world!?")
What is up world!?
```

In layman terms the print statement has been replaced by the `print()` function. If you're making a switch from Python2 to Python3, then one way to ease the transition is to start using the `print()` function. The good news is if you're converting a project from Python2 to Python3, and if they're many print statements in your code then it's mostly a nonissue when using the 2-to-3 code translation tool: http://bit.ly/2DquVkz

Control flow

Python like many programming languages allows programmers to control the order in which statements are executed via **control flow.** Python enables `if`, `elif`, and `else` statements. There's no support for `switch` statements in Python3, and the logic can instead be implemented using chained `elif` statements. The following show how these statements work...

```python
if 5 < 10:
    print("CORRECT!")
...
CORRECT!
```

In the above code snippet the condition to be tested is 5 < 10. If you're coming from a C-style language like Java then this syntax takes some acclimation. For example, in Java an `if` statement would look something like:

```java
if(5 > 3) {

    System.out.print("CORRECT");
}
```

However, in Python, parentheses are optional and the expression ends where the colon (:) is at. Also, the curly braces are not needed but instead four spaces are recommended to represent the body of the condition. The following code snippet represents an if-else statement in Python:

```python
if 5 > 10:
    print("CORRECT")
else:
    print("WRONG")
...
WRONG
```

It's similar to the if statement except that it also has an else statement which is executed by default if the text expression evaluates to false. The last flavor of if statements are elif. There can be 0 or more elif statements and the else statement is completely optional. The elif keyword is shorthand for 'else if'. Multiple elif statements can be chained together to create the logic that's equivalent to a switch statement. The below code snippet illustrates elif statements in Python:

```python
if 1 > 2:
    print("1 is greater than 2")
elif 1 < 2:
    print("1 is less than 2")
elif 1 == 2:
    print("1 is equal to 2")
else:
    print("Something's fishy")
...
1 is less than 2
```

The above code snippet has an `if` statement, two `elif` statements, and a single `else` statement. The `if` statement is checked first, and if `true` is executed; if not, then the first `elif` statement is checked, and if it evaluates to `false` then the second `elif` statement is checked. If none of the above three statements evaluates to `true` then then the `else` statement is executed by default.

Iteration in Python

Iteration can be defined as a repetitive process, a cycle if you will. Iteration is something that computers don't mind while humans will find this type of work humdrum. There are two ways to perform iteration in Python: while loops, and for loops. At first glance the differences between them seem trivial but with experience their distinctness becomes clear.

While loop

In plain English, the while loop states "that while something is true, then do this." A simple example from the real world is *while its business hours, keep the store open*. Now, let's translate this logic into Python pseudo code:

```
while condition is True:
    do this
```

There are a couple of things to keep in mind *while* doing this (pun was seriously not intended)

- The `while` keyword is mandatory. It's a reserved keyword in Python and excluding it will cause the compiler to flip-out.
- The colon is needed after the condition.

- It's suggested to use **four spaces per indentation level**. This is used largely by the Python community, and using a tab which is about 8-spaces is discouraged.
- You could use curly braces to separate the blocks of code, but it's completely optional and will make your code look out-of-sync to other Python code.
- Parenthesis that encloses the condition is legal but not recommended. Therefore, the following statement is also legal:

```
while(x < 5):
```

Below are some examples that show how to use while loops in Python.

```
x = 1
while x < 5:
    print("x = ", x)
    x = x + 1

...


x = 1
x = 2
x = 3
x = 4
```

In the above code x is set to 1, and is commonly referred to as the **counter variable**. The reason for this is because it counts up to the upper limit from its starting position. The condition of the loop is x < 5, so the loop will increment until x is less than 5 which is when x equals 4; the contents of x is then printed to the console. The counter must be updated or less the loop won't stop and then you'll be left with what's known as an **infinite loop** which sucks. In this example, the loop is updated using x = x + 1. However, shorthand notation could alternatively be used such as x+=1 which is logically equivalent to x = x + 1. Next, let's use a while loop to print the first 100 Fibonacci numbers. For a quick refresher, Fibonacci numbers are a special sequence which states that every number after the first two are the sum of the two preceding ones.

Here's a list of some of the Fibonacci numbers:

1, 1, 2, 3, 5, 8, 13, 21, 34, 55, 89, 144, …

Here's the Python code that will compute the first 100 Fibonacci numbers:

```
n, a, b, count = 0, 1, 1, 1
while n <= 99:
   print("n =", count, ":", a)
   a, b = b, a + b
   n+=1
   count+=1
```

Here's an alternative sequence for Fibonacci numbers:

0, 1, 1, 2, 3, 5, 8, 13, 21, 34, 55, 89, 144, …

Here's the corresponding while loop for this:

```
n, a, b = 0, 0, 1
while n <= 100:
   print("n =",n, ":", a)
   a, b = b, a + b
   n+=1
```

Fibonacci numbers can get big in a hurry but good thing modern CPUs are equipped to handle gargantuan numbers.

The 100th Fibonacci number is monstrous; it's equal to:

354224848179261915075

The for loop

The for loop in Python is a wee-bit different from the for loop in C-style languages such as Java that allows you to set a starting variable, condition, and then update that starting variable. Instead, Python will allow you to iterate over a sequence of numbers using the built in range() function. The pseudo code below illustrates the structure of a for loop in Python:

```
for variable in range():
    do this
```

Notes about the for loop:

- for is a reserved keyword in Python so it can't be used as a variable name.
- The condition must follow the for keyword.
- A colon (:) must immediately follow the end of the condition.
- It's recommended to indent the statements that goes inside the body of a for loop by four spaces, like that of a while loop.
- The for loop can be used to iterate over the built-in data structures in Python like strings, lists, sets, tuples, and dictionaries.

Here's a simple example of how the for loop works in Python:

```
for x in range(5):
```

```
print("x =", x)
...

x = 0
x = 1
x = 2
x = 3
x = 4
```

With the range() function the starting point is at 0 or less specified, and the ending point is at 5. Note, the range() function will never print the endpoint, it will always stop one short of it. So, the starting point is inclusive and the ending point is exclusive. It's important to note that the range() function does something a little funky when you try and print the values that's inside a prospective range. For example, look at the following:

```
>>> print(range(1,10))
range(1, 10)
```

So, yeah, there's no shortcut for generating sequences via the for loop in Python. Here's another example of the for loop in action:

```
for x in range(50, 100):
    if x % 2 == 0 and x % 3 == 0:
        print("ABC")


...
ABC
ABC
ABC
ABC
ABC
```

ABC
ABC
ABC

The above makes use of the **modulus operator** (%) which checks the remainder after it's divided by a number. For example, since 50/ 2 equals 25, there's no remainder. The loop check what numbers from 50 to 99 is divisible by 2 AND 3. If a match is found then the text "ABC" is printed. You can also specify the "step" in which you'll iterate over the sequence by adding a third parameter. For example, let's look at the following code snippet:

```
for x in range(1,10,2):
    print(x)
...
```

The output when printed is:

```
1
3
5
7
9
```

The loop starts at 1 and goes to 10. The step in which it iterates is 2, so the loop prints 1, 3, 5, 7, 9. If you want to print all of the even numbers in the range of 0-10 including 10 then you'll write this:

```
for x in range(0,11,2):
    print(x)
...
```

```
0
```

```
2
4
6
8
10
```

You can **nest** loops as many levels as you wish, or including a loop inside another. For example, here's a triple loop, or two loops inside another loop:

```python
for i in range(1,10):
    for j in range(1,3):
        for k in range(1,3):
            print("i=", i, "j=", j, "k=", k)
```

The output is this:

```
i= 1 j= 1 k= 1
i= 1 j= 1 k= 2
i= 1 j= 2 k= 1
i= 1 j= 2 k= 2
i= 2 j= 1 k= 1
    .

    .

    .

i= 9 j=2 k=2
```

The *triply* nested loop works in a similar fashion to the doubly nested loop. The outer loop controls the inner loop, and that loop controls its inner loop. Therefore, the inner loops will iterate at a faster rate. This loop iterates a total of 36 times, which can be computed by taking the bounds of each loop and multiplying them by each other, aka 9 x 2 x 2.

break, continue, and pass statements

Imagine that a loop you write is a contract. What you think happens if you broke the contract of the loop? One thought would be that it would be terminated. You can definitely terminate a loop by breaking it and how to do so is listed below:

```
x = 0
while(True):
    print(x)
    if(x == 1000):
        break
    else:
        x = x + 1
```

The above prints the numbers from 0 to 1000 and *breaks* or terminates at 1000. This is important because the condition of `while(True)` means that the loop will iterate indefinitely because that condition is always True. Therefore, a way to exit out of the loop is to inject a conditional that triggers the end of a loop. The continue statement does the opposite. It skips the body statements once the condition is True:

```
while(x < 10):
    x+=1
    if(x == 5):
        continue
    else:
        print(x)
```

```
1
2
3
4
6
7
8
9
10
```

Also, there's the pass statement which is simply a placeholder. You would use it when a statement is required syntactically but no code needs to be executed. At first glance this seems identical to the continue statement but to get a better understanding let's look at the two snippets of code:

Code a:

```
for x in range(1,10):
   if(x == 5):
        pass
        print(x)
   else:
        print(x)
...
1
2
3
4
5
6
7
```

8

9

Code b:

```
for x in range(1,10):
    if(x == 5):
        continue
        print(x)
    else:
        print(x)
```

...

1

2

3

4

6

7

8

9

In code snippet a, the pass statement does nothing so the values of 1-9 are printed. In code snippet b, the continue statement is instead used which forces execution of the next iteration. Therefore, the statements following the continue statement are not executed and the numbers 1-4, and then 6-9 are printed.

Iterating over data structures in Python

It didn't make sense to teach how to iterate over the built-in data structures without first showing the common loops – that would've been bad teaching. Now that you've gotten acquainted with them it's time to learn how to iterate over lists in Python. Let's take a look at some examples.

```
>>> animals =  ["lion", "fossa", "okapi", "spider crab",
"maned wolf"]
>>> for x in animals:
        print(x)
...
lion
fossa
okapi
spider crab
maned wolf
```

The following code snippet checks to see if "okapi" is in the list and if it is it gets printed:

```
animals = ["lion", "fossa", "okapi", "spider crab",
"maned wolf"]

for x in animals:
  if x == "okapi":
      print(x)
...
okapi
```

You could alternatively use the while loop to iterate over lists. You'll need the help of the built-in len() function so that you know where to increment up to. Below is an example of how you'll accomplish this with a while loop:

```
animals =  ["lion", "fossa", "okapi", "spider crab",
"maned wolf"]

i, j = len(animals), 0
while j < i:
  print("index = ", j, ":", animals[j])
```

```
   j+=1
...
index = 0 : lion
index = 1 : fossa
index = 2 : okapi
index = 3 : spider crab
index = 4 : maned wolf
```

Lists can also be nested or one included inside another one. For example, take a look at the following:

```
>>> nest = [["Hello", "Jambo", "Hola", "Namaste",
"Salaam"], ['a', 'b', 'c', 'd'], [123, 3302,
2.2120, -2002, .292010]]
```

This is a list that itself contains three elements which are lists. To access the elements of the list you can use the following notation list[element]. Therefore, the following will give this:

```
>>> nest[0]
['Hello', 'Jambo', 'Hola', 'Namaste', 'Salaam']
>>> nest[1]
['a', 'b', 'c', 'd']
>>> nest[2]
[123, 3302, 2.212, -2002, 0.29201]
```

However, what if you want to access "Jambo" inside of the first list? Since you have a list inside a list, you must use double subscript notation which looks like:

```
nest[list][element]
```

Therefore, to access "Jambo" you'll need to do the following:

```
>>> nest[0][1]
'Jambo'
```

If you want to go a step further then you don't have to just stop at double nested lists, you can create triple nested lists, or two lists nested inside a single list. Let's take the following as an example:

```
>>> triple = [['a', 'b', 'c'],"Hola", [1, 2, 3, 4,
5, ["aa", "bb", "cc"]]]
>>> triple[0]
['a', 'b', 'c']
>>> triple[1]
'Hola'
>>> triple[2]
[1, 2, 3, 4, 5, ['aa', 'bb', 'cc']]
>>> triple[2][5]
['aa', 'bb', 'cc']
>>> triple[2][5][0]
'aa'
```

The above list named triple contains three elements. There's a list, a string, and another list which itself contains another list. To access an element in the triply nested list you must use the following notation:

```
listname[outerlist][innerlist][element]
```

In other words you'll have to access the outer list or the main list, then the inner list which contains the element, and then the index at where the list is located. So, to access the string "cc" you'll have to do:

```
triple[2][5][2]
```

If you want to iterate through nested lists, then you can use nested loops or loops inside another loop. Let's first take a look at nested loops so that we can gain a better understanding of how they work. Look at the following code snippet:

```
for x in range(1, 13):
    for y in range(1, 13):
        print(x, "x", y, "=", x * y)
        if y == 12:
            print("------------")
```

The following prints the multiplication tables of 1-12. It prints a dotted line after every multiplication table so that the output is easier to read. How nested loops work is that the outer loop controls the inner loop. Therefore, once the outer loop starts, the control then passes to the inner loop which increments until the given condition evaluates to false.

Iterating over Nested lists

Once you know how nested loops work then iterating over nested lists becomes straightforward. The following code fragment shows how to iterate over nested lists:

```
>>> us_cities = [["Los Angeles", "Seattle",
  "Portland"],["NYC", "Philadelphia","Boston", "Raleigh",
  "Atlanta", "Miami"]]

for x in us_cities:
    for y in x:
        print(y)
...
```

Iterating over tuples is similar to iterating over lists as shown below:

```python
nums = (0, 25, 78, 837, 982)
for x in nums:
    print(x)
...
0
25
78
837
982
```

You can use the `index()` function to get the number of each index as shown below:

```python
for x in nums:
    print(nums.index(x), x)
...
0 0
1 25
2 78
3 837
4 982
```

You can also add a counter to an iterable by using the `enumerate()` function:

```python
for x in enumerate(nums):
    print(x)
...
(0, 0)
(1, 25)
```

```
(2, 78)
(3, 837)
(4, 982)
```

To iterate through a dictionary in Python you can use the
`items()` function:

```
quote = {

  "a": "hahahahahah",
  "b": "bwahahahahaha",
  "c": "cool beans"
}

for i, j in quote.items():
  print(i, j)


...
a hahahahahah
b bwahahahahaha
c cool beans
```

You can also iterate over strings the same way you would a
list:

```
s  = ""
vowels = "aeiou"
for string in vowels:
  s += string
...
>>> s
'aeiou'
```

Creating Functions

A function in mathematics is similar to a function in computing. You take in some input and then it spits out some output. Here's a list of some simple functions in mathematics:

- Square function: $f(x) = x^2$
- Cube function: $f(x) = x^3$
- Square root function: $f(x) = \sqrt{x}$
- Reciprocal function: $f(x) = \frac{1}{x}$

To create a function use the following syntax:

```
def function_name(parameters)
    insert statements
```

The word def is a reserved keyword which you can use to represent that you have created your function. The return keyword is a reserved keyword that reveals that the function returns a value when called. Below is a list of simple functions in mathematics converted to Python:

```
def square(n):
   return n * n
>>> square(2)
4

def triple(n):
      return n * n * n
>>> triple(4)
64
```

```python
from math import sqrt
def square_root(n):
    return sqrt(n)
>>> square_root(28)
5.291502622129181


def reciprocal(n):
            return 1/n
>>> reciprocal(5)
0.2
```

Once you create a function you need a way to use it which in technical terms is called *invoking a function*. There are several ways in which you can invoke a function in Python. A simple way as you saw from the previous example is to include the function name and then pass the input aka **arguments** into the parentheses. Therefore, it'll look something like the following:

```python
c = function(a,b)
```

You can also call functions using keyword arguments. When the function is invoked a specific keyword or words are used to specify the input into the function. For example, the function looks like the following:

```python
def sounds(goat="bleat", mice="squeak", oxen="moo",
horses="neigh"):
    print("Billy goats", goat)
    print("Mice", mice)
    print("Ox", oxen)
    print("Horses", horses)

>>> sounds("neigh", "oink, oink", "talk", "bark")
```

```
Billy goats neigh
Mice oink, oink
Ox talk
Horses bark
```

Python also includes two helpful conventions when it comes to creating functions which are the * and ** symbols. When a single star is used then you can pass in an arbitrary number of arguments into the function. Even though args is typically used as the parameter, it can in fact be any name. An example of the function in action is listed below:

```
def fun(*args):
    for x in args:
        print(x)
...
>>> fun(1, 10, 20, 30, 40, "Hippo!")

1
10
20
30
40
Hippo!
```

When you include two stars this means that you can use an arbitrary number of **keyword arguments** or **kwargs for short. Like with args, kwargs is simply a variable and not a reserved keyword. Below is an example of how to use kwargs:

```
def function(**kwargs):
    for key, value in kwargs.items():
        print(key, ":", value)
...
```

```
>>> function(name = "Sir James", city = "Tokyo",
fruit = "Apple", animal = "Bear")

name : Sir James
city : Tokyo
fruit : Apple
animal : Bear
```

Functional Programming in Python

In programming a **lambda** is a function that's not glued to an identifier. Hence, it's commonly referred to as an *anonymous function*. Lambdas are commonly used in functional programming languages like JavaScript and Scala. Let's first look at a regular function and then we'll translate that into an anonymous one.

```
def algebra(x, y, z):
    return 5*x + 6*y - z
...
>>> algebra(1, 2, 3)
14
```

The above function is a simple mathematical equation that takes in some inputs and returns a value. We can translate this into an anonymous function as follow:

```
>>> result = lambda x,y,z : 5*x + 6*y - z
>>> result(1,2,3)
14
```

You could compute the above without assigning it to a variable but the issue with that is that you'll never be able to use the anonymous function again. That's OK if you have plans on using it in a one-off fashion, but if not then assigning it to a variable is a good idea for reuse. Here's the general syntax for lambda functions:

```
lambda input : output
```

The lambda keyword is a reserved one in Python and is followed by all of the input variables. A colon (:) is used to separate the input from the output, or what's returned. Lambda expressions are semantically equivalent to classical functions it's just that the syntax is a little off – for example, a lambda function does not use the def or return keywords like classical functions do. Lambda functions are to only be used for short-and-sweet expressions as multiline lambdas are not permitted.

There are also several other cool things that you can do with the aid of lambdas such as map, filter, and reduce. For example, let's analyze the following code fragment:

```
numbers = [1,2,3,4]
list = [ ]
for x in numbers:
    list.append(x*3)
>>> list
[3, 6, 9, 12]
```

The following code can be replicated using what's known as a **map** shown below:

```
>>> numbers = [1,2,3,4]
>>> list(map(lambda x: x*3, numbers))
[3, 6, 9, 12]
```

The lambda expression is passed to the map() function

which returns an iterator object that applies the function to every item of the iterable. The returned result is then passed to the list() function that converts the output to a list. In various programming languages a map is considered a **higher-order** function that applies a given function to every element in a list. A **filter** is a function that extracts each element of the sequence for which the function returns True. Look at how the filter is applied to the following function:

```
>>> data = [.0212, .789, .897, .9821, 1.020, 1.121,
1.567]
>>> list(filter(lambda x: x > 1, data))
[1.02, 1.121, 1.567]
```

The lambda accepts an input of x and returns x > 1. A list named data is supplied as a second argument in the lambda and then the filter function is applied to all of the elements in data; only the elements that are greater than 1 are returned. The **reduce** higher order function takes a list and then reduces it to a single value by applying the function. In Python3 it's not a built in function, but it has been sent to the functools module. Guido van Rossum, the creator of Python suggests that *an explicit for loop is more readable than functools.reduce() 99% of the time*. For an example, let's say that you have a list of numbers and simply wanted to sum all of them up. One way to do this is as follows:

```
data = [1,1,2,2,3,4,5]
i = 0
for x in data:
   i+=x
>>> i
18
```

However, you can make your code more compact by using the reduce() function. First, import reduce() from the functools module and then pass the lambda into reduce() as shown below:

```
>>> from functools import reduce
>>> data = [1,1,2,2,3,4,5]
>>> reduce(lambda x, y: x + y, data)
18
```

There's also something called **list comprehensions** which are a syntactic construct that allows for the creation of new lists based on existing ones. For example, let's say that we want to create a list of booleans that returns True or False for the first 10 positive digits. We want to return True if the digit is greater than 5, and False any other time. One way to do this would be as follows:

```
truth = []
for i in range(1,11):
    if i > 5:
        truth.append("True")
    else:
        truth.append("False")
```

```
>>> truth
['False', 'False', 'False', 'False', 'False',
'True', 'True', 'True', 'True', 'True']
```

However, with list comprehensions we can get the same result by using just a single line of code:

```
>>> [x > 5 for x in range(1,11)]
[False, False, False, False, False, True, True,
```

True, True, True]

Recursion

Let's explore a classical mathematical problem which is how to compute factorials. The equation for it is described below:

n! = n *(n-1)* (n-2) … 3 * 2 *1

Therefore, we can compute the factorial of 4 by doing the following:

4 * 3 * 2 * 1 = 24

You could write a function that solves this iteratively as shown below:

```
def factorial(n):
    fact = n
    count = fact - 1
    while count > 0:
        fact*=count
        count-=1
    return fact

>>> factorial(10)
3628800
```

You could alternatively reverse the direction in which you're iterating from and instead of top-to-bottom go from bottom-to-top:

```
def fib(n):
    fib, count = 1, 1
```

```
    while count <= n:
        fib*=count
        count+=1
    return fib

>>> fib(10)
3628800
```

However, there's another way to compute Fibonacci numbers which is by using recursion – A **recursive function** is a function that calls itself. There must be what's known as a **base case** because that makes the function terminate or else the code runs indefinitely... bummers. Below is an implementation of recursion in Python:

```
def fact(n):
    if n == 1:
        return 1
    return n * fact(n-1)

>>> fact(8)
40320
```

As you can see the base case n == 1 check to see if the result is equal to 1; if so, then it returns 1 and the function stops calling itself. If not, then the function continuously calls itself due to this part:

```
return n * fact(n-1)
```

To gain a deeper understanding of factorials in your spare time you could research some fun topics like *Droste effect*, cool artwork like *Fraser spiral, or even* interesting veggies like the *Romanesco broccoli*.

Error handling with Python

Errors are common in programming and anyone that says otherwise is a bonafide liar ;). Since errors are inevitable it's not a bad idea to have a mechanism in place to detect them as life will be a bit easier. One of these protection mechanisms are known as **exceptions** which help make your code more secure. A cardinal sin in mathematics is to *divide by 0;* it's so *notorious* that it's like trying to sell booze during Prohibition. This practice is also illegal in Python as shown in the following code snippet:

```
>>> 5/0
Traceback (most recent call last):
File "<stdin>", line 1, in <module>
ZeroDivisionError: division by zero
```

Aha, as you can see the compiler provides a detailed message which says:

```
ZeroDivisionError: division by zero.
```

You can process this error by using a try-and-except statement shown below:

```
try:
    5/0
except ZeroDivisionError:
    print("can't divide by zero my friend!"
...
can't divide by zero my friend!
```

There are many types of exceptions in Python. You can learn more about them from the Python docs:

http://bit.ly/2KbTWRS

Look at the following code fragment:

```
>>> 1 + "two"
Traceback (most recent call last):
File "<stdin>", line 1, in <module>
TypeError: unsupported operand type(s) for +: 'int'
and 'str'
```

The exception thrown is TypeError and arises when you're trying to add conflicting types. The way to handle exceptions in Python is with the following template:

```
try:
    block 1
except Exception1:
    trigger-exception1
except Exception2:
    trigger-exception2
else:
    else block
finally:
    final block
```

The try block is the portion of code that's attempted. Just think of it as saying *I'm going to try and accomplish something!* The except block is the portion of code in which specific errors are caught. There can be one or more except blocks in the code. If an exception is of type Exception1 then trigger-exception1 executes, and if an exception is of type Exception2 then trigger-exception2 is ran. If none of the exceptions are triggered then the code in the else block is executed. The finally block is the portion of code that's executed at the end. This block of code will run regardless and is typically reserved for cleaning up resources. Here's a try-and-except statement that handles 1 + "two."

```
try:
    1 + "two"
except TypeError:
    print("can't add different types!")
finally:
    print(1, "+", "two", "should be: 1 + 2 = ", 1 + 2)
```

...

```
can't add different types!

1 + two should be: 1 + 2 = 3
```

Classes in Python

Python allows programmers to do object oriented programming which will be explained in depth in the next chapter. However, a fundamental component of OOP is **classes** which are considered blueprints for building objects. To create a class in Python use the following syntax:

```
class Name:
    pass
```

This is literally the simplest class that you can create in Python; the pass placeholder which translates to *do nothing* is needed in order for the program to not throw errors. The keyword class is a reserved one in Python, and by convention class names are capitalized. Let's look at some code because in this case a little code is better than a 2-page explanation. Below is an example of a class in Python that computes all of the factors of a number:

```python
import math
class Factors:
    """computes the factors of a number"""
    def __init__(self, num):
        self.num = num
        print("computing factors...")
    def compute_factors(self):
        i = 2
        factors = []
        upperbound = math.floor(self.num / 2)
        while i <= upperbound:
            if self.num % i == 0:
                factors.append(i)
            i+=1
        return factors
```

```
>>> factors = Factors(100)
computing factors...
>>> factors.compute_factors()
[2, 4, 5, 10, 20, 25, 50]
```

Classes are typically seen in object oriented programming languages as an instance of the class – this instance creates an object which contains the data of the class. This object can then be used to call the methods in a class, aka a function that's binded to an object. If all of this doesn't quite make sense yet then no worries as these confusions will be ironed out in the object oriented programming chapter.

Modules

A **module** is a file that contains Python's definitions and statements. This will be important to know because after a certain time in coding you'll probably want your code to be saved. This is not the case when using the Python Shell as once you exit all the code inside of it is *poof...* gone. Therefore, if you want to reuse code one option is to place it inside a module. This process is super simple so let's walk through an example:

1. Open up a text editor and type the following:

```python
def sequence(start, n):
    i = 0
    series = []
    if start == 0:
        return 1
    while start < n:
        if (start % 2 == 1):
            series.append(start + start * 2)
            start += 1
        else:
            series.append(start - 1)
            start += 1
    return series
```

2. Save the file as: Sequence.py
3. Open up the terminal and change into the directory where Sequence.py is located. Next, open the Python interpreter.
4. Type the following:
```
>>> Sequence.sequence(5,20)
[15, 5, 21, 7, 27, 9, 33, 11, 39, 13, 45, 15, 51, 17, 57]
```

The code snippet in the module contains a function named sequence() which contains two arguments: start, and n. The function iterates up to n, and if start is divisible by 2, then start + start x 2 is added to the list; if not, 1 is subtracted from start and added to the list. There are many variations in how you can import a module. You could rename the filename to something more concise to prevent less typing such as:

```
>>> import Sequence as s
```

You could import specific functions from a module.

```
>>> from Sequence import sequence
>>> sequence(9.75,20)
[8.75, 9.75, 10.75, 11.75, 12.75, 13.75, 14.75,
15.75, 16.75, 17.75, 18.75]
```

Packages

A **package** is a way of organizing module names in Python. This is useful if you have many related modules and you want to create a logical hierarchy to organize the files. So, in other words, packages are no different than directories but with one special twist which is that each package must contain a file called __init__.py. This file which could be empty signals that the directory is a Python package; this file could also be imported like a regular module. For example, let's assume that we're creating a retro video game in Python and that we have the following files:

```
game/
__init__.py
draw.py
sound.py
color.py
```

If we want to access *draw*.py of the game package then you could use *dot notation*, so in this case:

>>> import game.draw

You can also use the from keyword to select individual files you want to import:

>>> from game import sound, color

Chapter 2 optimized

This chapter provided a high level overview of the basics of Python programming. You learned about the essential components of building Python programs such as variables, data structures, control flow, iteration, classes, modules, and many more. Instead of breaking these components into distinct chapters I've decided to combine them all into a single chapter so that you can get the basics out of the way and move to more interesting topics.

Gaining a solid understanding of these principles will make diving into more complicated Python material a smoother transition. I would highly recommend completing the chapter II challenges before proceeding. It's a good idea to give the problems your best effort before looking at the solutions as *sweating it out* is where the real learning takes place. You can find the solutions to this chapter's coding challenges here: http://bit.ly/2RAky2b

Chapter 2 Coding Challenges

Challenge 1: Write a function named `factorial_reduce()` that computes the factorial of any positive number with the caveat being that you must use a lambda and the `reduce()` function from the `functools` package. The function should be able to compute the factorial in just a single line. Here's the template of the code to get started:

```
from functools import reduce
def factorial_reduce(n):
    return...
```

Challenge 2: Write a function named factors() that lists all of the factors of an integer n. Below is the function signature:

```
def factors(n):
    """lists the facts of an integer n"""
```

Here are some sample test cases so that you can test your code against:

```
>>> factors(7)
1
7
```

```
>>> factors(14)
1
2
7
14
```

```
>>> factors(20)
1
2
4
5
10
20
```

```
>>> factors(2.1)
Traceback (most recent call last):
File "<stdin>", line 1, in <module>
File "<stdin>", line 3, in factors
TypeError: 'float' object cannot be interpreted as
an integer
```

Challenge 3: Write a function named
fo_shizzle_my_nizzle() that prints a string contingent on
the following conditions. The letter n is used as input.

fo → n is less than 0.
shizzle → n is in the range of 1-49, 1 is inclusive, and 49 is
exclusive.
my → n is in the range 50-100, with 50 and 100 being
inclusive.
nizzle → n is even, divisible by 3, and greater than 100.
" "→ if none of the above conditions are met then an empty
string is printed by default.

Write a loop that tests the input in the range of -10...150.
Here's some sample test cases:

```
-5 → 'fo'
0 → ' '
8 → 'shizzle'
```

```
52 → 'my'
150 → 'nizzle'
```

Challenge 4: Write a function called `compute_pattern()` that accepts a number n and then prints the following pattern:

```
1
2 3
3 4 5
4 5 6 7
5 6 7 8 9
6 7 8 9 10 11
7 8 9 10 11 12 13
8 9 10 11 12 13 14 15
9 10 11 12 13 14 15 16 17
```

Challenge 5: Bubble sort is classic sorting algorithm. It's not recommended to use when performance is important, but regardless of its shortcomings it's still a nice learning exercise. Write a function called `def bubble_sort()` that accepts a list named `arr`, and then sorts it from least to greatest order. Below is a sample list named items to test on:

```
items = [92, 7, 38, 37, 92, 37, 12, 54, 43, 67, 78,
83, 93, 101, 128, 139, 156]
```

Here's the output:

```
>>> bubble_sort(items)
[7, 12, 37, 37, 38, 43, 54, 67, 78, 83, 92, 92, 93,
101, 128, 139, 156]
```

Chapter III: Object oriented programming in Python

fdecomite Follow - Some new Escher Cookie Cutters - Image - CC BY 2.0

Kind of like there are many different styles to fashion, there are also various styles to coding such as imperative, functional, and object oriented. With object oriented programming (OOP) data is wrapped inside something known as an *object*. The object contains the data which can be accessed when needed. OOP is a popular coding paradigm that's used by many software companies and a reason for this is that it helps make code more modular, facilitates code reuse, and makes the code base more maintainable. In this chapter you'll learn how to do OOP programming in Python along with popular concepts such as classes, methods, inheritance, and polymorphism.

A simple class to explain basic OOP principles

Let's get our hands dirty and look at some code to gain a better example of how OOP works:

```python
class Equations:
    """this class holds various mathematical equations."""
    def polynomials(self,x,y,z):
        return 5*x + 10*y + 2.5*z


...


>>> a1 = Equations()
>>> a1.polynomials(2,3,4)
50.0
```

Believe it or not, there's actually a lot happening in the above code snippet. One, the class is created with the class keyword and the class name must immediately follow the colon. Following the class name is the **docstring** which is a string that occurs as the first statement in a module, function, class, or method definition. The docstring is enclosed within triple quotes and lists the details about the class, and should be indented like a regular statement in Python. You can find out more details about the docstring convention in PEP 257: http://bit.ly/2zdWUku

To create an instance of copy of the class use the following notation:

```python
reference_variable = ClassName()
```

Therefore, the following statements create several instances of the Equations class.

```
a = Equations()
b = Equations()
c = Equations()
d = Equations()
e = Equations()
```

All of the above instances are of type Equations. To double check it here's the following statement:

```
>>> type(a)
<class '__main__.Equations'>
```

OK, so from the above example we've created two objects with the reference variable named a and b respectively. We have invoked the polynomials() method and both objects holds the same values. So, here's a quick quiz for you. What do you think the following code will print?

```
if a == b:
    print("Equal")
else:
    print("Not Equal")
```

. . .

```
Not Equal
```

I know it's kinda tricky. Since these are two separate objects they're not equal. When you're using the == operator to compare objects you're comparing their addresses in memory, not the value that they contain. For example, if you just type a or b in the terminal then you'll get something like the following:

```
>>> a
<__main__.Equations object at 0x7f0bbbb83b38>
>>> b
<__main__.Equations object at 0x7f0bbb4cf2b0>
```

If you want to compare their values then one way you can do this is by the following code snippet:

```
a1 = a.polynomials(1,2,3)
b1 = a.polynomials(1,2,3)
if a1 == b1:
    print("Equal")
else:
    print("Not Equal")
```

The following will output: Equal

The reason being is because you're comparing their values not their addresses.

The Self Variable

The self convention is a concept used in Python to represent an object. Since a class could have many objects, a convention to help keep track of them is useful. It's important to know that self is a variable, not a keyword, so you could theoretically use another variable name instead. However, according to PEP 8 it's always recommended to use self as other Python programmers will quickly understand what's happening. The self convention in Python is similar to the this keyword in C++ or Java. This concept is better explained with some code.

```python
class Humans:
    """Simple class for modeling humans."""
    def first_name(first):
        return first
    def last_name(last):
        return last


...


>>> name = Humans()
>>> name.first_name("Mike")
```

Here's the output:

```
Traceback (most recent call last):
File "<stdin>", line 1, in <module>
TypeError: first_name() takes 1 positional argument
but 2 were given
```

What the heck? The error doesn't make any sense because it says takes 1 positional argument but 2 were given. This compiler must be smoking some serious binary stuff because only one argument was provided, yet it claims two were. Let's try something, let's update the code so it includes the self keyword.

```python
class Humans:
    """Simple class for modeling humans."""
    def first_name(self, first):
        return first
    def last_name(self, last):
        return last
```

```
...
name = Humans()
name.first_name("Mike")
'Mike'
name.last_name("Capone")
'Capone'
```

OK, so the code now works. As you can see the way to fix this was to add self as the first parameter in all of the methods. To *call* or use a method you must use what's known as **dot notation**. This is a style that's popular in object oriented programming (OOP) languages and the dot means that it accesses a member of the class. You could also use the following notation to call a method of a class:

```
ClassName().method_name()
```

A concrete example is listed below:

```
>>> Humans().first_name("Maria")
'Maria'
```

There's some stuff happening behind the scenes. For example, when the method is called this is what's going on:

```
class Humans:
    """Simple class for modeling humans."""
    def first_name(jane, first):
        return first
    def last_name(jane, last):
        return last
```

The object that's being referenced which is jane replaces self so that the compiler knows what object it's working with. The self variable doesn't do anything except just refer to an object, in this case jane.

The __init()__ method

There's a special method in Python that allows objects to be created with an initial state. This method is known as __init()__ which is short for *initialization*. This is similar to the *constructor* in other OOP languages like Java. Let's use it to update the previous class so that we can create variables that can be used across the instance of a class. An example of it is listed below:

```python
class Humans:
    """Simple class for modeling humans."""

    def __init__(self, first, last):
        self.first = first
        self.last = last

    def first_name(self):
        return self.first

    def last_name(self):
        return self.last

    def full_name(self):
        return self.first + " " + self.last
>>> person = Humans("John", "Q")
>>> person.first_name()
'John'
>>> person.last_name()
'Q'
```

```
>>> person.full_name()
'John Q'
```

A couple of modifications were made. One, the __init__() method was added which includes the parameters of self, first, and last. Inside the __init__() method the values of __init__() are initialized with the following statements:

```
self.first = first
self.last = last
```

The statements mean to set the object's variable of first equal to first, and the object's variable of last equal to the value of last. Remember, the self variable is needed so that we can keep track of the object we're referencing. The rules of object oriented programming dictates that we can manipulate, add, delete, or access the data of a class. For example, we can add a new field by using the dot notation as follows:

```
person.occupation = "mechanic"
```

To access the field use the dot notation again:

```
>>> person.occupation
'mechanic'
```

To delete an attribute use the del statement:

```
>>> del person.occupation
```

You know it works as when you try to access the attribute it creates an error:

```
>>> person.occupation
Traceback (most recent call last):
File "<stdin>", line 1, in <module>
AttributeError: 'Humans' object has no attribute
'occupation'
```

You can also use the dot operator to modify the value of an attribute as shown below:

```
>>> person.first = "Johnny"
>>> person.first
'Johnny'
```

Also, another way you could of implemented the full_name() method is as follow:

```
def full_name(self):
    return self.first_name() + " " + self.last_name()
```

This uses the self variable again to access the value of the method that's associated with the object. Now that we have a general idea on how OOP works in Python we can start designing some simple programs the OOP way. For example, let's design a program that models a Point in mathematics. One solution is listed below:

```
class Point:

    """Models a point in mathematics."""
    def __init__(self,x,y):
        self.x = x
        self.y = y

    def getX(self):
        return self.x
```

```
    def getY(self):
        return self.y

    def get_point(self):
        point = (self.x, self.y)
        return point
>>> p1 = Point(1,2)
>>> p1.getX()
1
>>> p1.getY()
2
>>> p1.get_point()
(1, 2)
```

Inheritance

Python allows inheritance which is a critical component of OOP. **Inheritance** is when a class receives the properties and methods of a class. Kind of like how a child inherits genes from their parents in the form of 23-chromosomes, subclasses or *child* classes can inherit attributes of a base class or *parent*. A simple example of inheritance is illustrated below:

```
class A:
    """I'm the base class.Bwahahaha."""
    pass


class B(A):
    """I'm the child class."""
```

```
    pass
```

In the above example two classes named A and B are
created. The two classes both contain the pass statements
so they effectively do squat. The base class is A, and class B,
the child class inherits from the parent class by using the
syntax:

```
ChildClass(BaseClass):
```

The above syntax is how you denote that inheritance is
taking place. The example simply illustrates how to
syntactically represent inheritance, but doesn't illustrate the
power of it. As stated earlier, with inheritance the child class
retains the attributes and methods of the parent class. So,
let's test this by looking at some code:

```
class A:
    """I'm the base class.Bwahahaha."""
    def message(self):
        print("A")
class B(A):
    """I'm the child class."""
    pass
>>> a1 = A()
>>> a1.message()
A
>>> b1 = B()
>>> b1.message()
```

Ok, some updates were made to class A. Instead of simply
containing a pass statement, it has a message() method
which prints A. The child class is identical to the way it was
before. An instance of A and B are created, and then the
methods are invoked. When you call a1.message(), A is

printed as expected. However, when you call b1.message(), A is also printed. This is somewhat unexpected as in the previous message there's no method in its class. However, remember that all of the attributes and methods of the parent class is inherited by the child class therefore B can call any method that A has, but the reverse can't be said. Also, Python supports **multiple inheritance**. So, multiple classes could extend the base class. Check out the following code:

```python
class A:
    """prints hello"""
    def message(self):
        print("Hello")
class B:
    """adds two numbers"""
    def add(self,x,y):
        print(x + y)
class C:
    """multiplies a number by 10"""
    def scales(self,num):
        print(num*10)
class D(A,B,C):
    """Class D extends A,B, and C"""
    def message(self):
        print("Hola")
        def read_input(self):
            message = input("Enter your message: " )
            return "Your response was: " + message


>>> d = D()
>>> d.message()
Hola
>>> d.add(5,10)
```

```
15
>>> d.scales(10)
100
>>> print(d.read_input())
Enter your message: Hey there
Your response was: Hey there
```

The above code has a total of four classes named A, B, C, and D. Each of the classes contains one method except for D which extends classes A, B, and C. However, let's observe class D more carefully. It has two methods in it which are message(), and read_input(). The message() method prints *Hola*, and the read_input() method reads in user input and returns the text. Here something simple and powerful is happening in class D. It contains the method message() which is different from the one in class A. This is known as **method overriding** which means that the functionality of the parent method is overridden in the child's method which is why Hola instead of Hello is printed.

Polymorphism

Poly means many, and *morphism* means change or form. Combined together and poly + morphism translates into *of many forms*. One of the benefits of polymorphism is that it allows coders to specify methods in an abstract or general way, and then implement them in particular instances. Think of animals like dogs, bees, cats, cows, and horses. They are all animals and make sounds, but their sounds are all different. Let's model this concept with some Python code:

```python
class Animal:
    def sound(self):
```

```
        raise NotImplementedError("subclass must
implement abstract method.")
class Dogs(Animal):
    def sound(self):
        pass
class Dogs(Animal):
    def sound(self):
        return "bark bark bark"
class Bees(Animal):
    def sound(self):
        return "buzz"
class Cats(Animal):
    def sound(self):
        return "meow"
class Horses(Animal):
    def sound(self):
        return "neigh"
class Cows(Animal):
    def sound(self):
        return "moo"

>>> d = Dogs()
>>> d.sound()
'bark bark bark'
>>> b = Bees()
>>> b.sound()
'buzz'
>>> c = Cats()
>>> c.sound()
'meow'
>>> h = Horses()
>>> h.sound()
'neigh'
>>> c = Cows()
```

```
>>> c.sound()
'moo'
```

The base class is Animal and it contains a method named sound(). All of the sub classes extend the Animal class and override the method of sound() to implement their own version of it.

Encapsulation

Encapsulation can be thought of as the art of *data hiding*. Data is very important, so you don't want it to get modified when it's not supposed to. Encapsulation restricts access to methods and variables which prevents the data from being modified by accident and therefore makes the code more *robust*. Let's model a popular object that we use often which is our account. A bank account holds data that we can use:

```
class BankAccount:
    """A simple bank account class"""
    def __init__(self, amount):
        self.balance = amount
```

```
>>> account = BankAccount(1000000)
>>> account.balance
1000000
```

The result is predictable right. That's because the variable is *public* meaning that it can be accessed anywhere. Well, let's change things up and do some modifications. It's possible to emulate *private* and *protected* variables and methods in Python. You'll need to use a process called **name mangling:**

```
class BankAccount:
    def __init__(self, amount):
        print("Inside the constructor")
        self.__balance = amount

account1 = BankAccount(10000)
account1.__balance
```

```
>>> account1.__balance
Traceback (most recent call last):
File "<stdin>", line 1, in <module>
AttributeError: 'BankAccount' object has no
attribute '__balance'
```

In this example a variable named __balance is created inside the constructor. When the constructor is called a message is printed and self.__balance is set to the value of amount. When a BankAccount object is made the value of 10000 is set to account1, and then account.__balance is accessed. The issue with this is that __balance is a private variable meaning that it can only be accessed by the class itself. The value of __balance can be accessed but the way you do this is by returning the value of the private variable through a method. An example of the updated code snippet is listed below:

```
class BankAccount:
    def __init__(self, amount):
        print("Inside the constructor")
        self.__balance = amount
    def getBalance(self):
        return self.__balance
```

```
>>> account1 = BankAccount(10000)
Inside the constructor
```

```
>>> account1.getBalance()
10000
```

You can also make variables protected in Python. This means that the variable could be accessed by the class itself, and sub classes. You denote a protected variable in Python by appending a single underscore at the beginning of the variable name as shown below in the following code snippet:

```python
class BankAccount:
    def __init__(self, amount):
        print("Inside the constructor")
        self.__balance = amount
    def getBalance(self):
        return self.__balance
```

Methods like variables can also be made private and protected, and just like their variable counterparts, you do this by modifying their name with double and single underscores respectively...

Private and Protected Methods in Python

A private method is one that can only be accessed from within the class itself. A protected method on the other hand can be accessed from the class and its subclasses. The following code snippet displays private and protected methods in Python:

```python
class BankAccount:
    def __init__(self):
```

```
        self.__balance = 0
        self.__account_num = 1234567891
    def __get_account_details(self):
        print(self.__balance)
        print(self.__account_num)
    def _change_account(self, balance, account_num):
        self.__balance = balance
        self.__account_num = account_nu

>>> account = BankAccount()
>>> account.__get_account_details()


Traceback (most recent call last):
File "<stdin>", line 1, in <module>
AttributeError: 'BankAccount' object has no
attribute '__get_account_details'
```

Calling a private method directly will result in an error. You can use the following syntax to call a private method:

```
>>> account._BankAccount__get_account_details()
10
1111111111
```

You can call a protected method directly by using the following syntax:

```
object._method()
```

Here's an example of it in action:

```
>>> account._change_account(7000, 2223334444)
7000
2223334444
```

The Super Keyword

According to the Python docs the super function *returns a proxy object that delegates method calls to a parent or sibling of a type*. This can be beneficial when you need to access inherited methods that have been overridden in a child class. It can also be useful in a case of single inheritance when the parent class needs to be accessed without being named explicitly. Let's look at some code:

```python
class ParentClass:
    """super() demo"""
    def __init__(self, x, y, z):
        self.x = x
        self.y = y
        self.z = z
        print(self.x, self.y, self.z)
class SubClass(ParentClass):
    def __init__(self, x, y):
        self.x = x
        self.y = y
        print(self.x, self.y)
        super().__init__(6,7,9)

>>> p = ParentClass(1,2,3)
1 2 3
>>> s = SubClass(4,5)
4 5
6 7 9
```

In this example there's a parent class along with its child class. The parent class contains a constructor which initializes three variables and then prints them. In the subclass there's also a constructor which initializes two variables and then prints them. The subclass makes a call to the parent class's constructor using the `super()` method, and three arguments are passed to it. The control flow goes to the parent class and then the code executes which results in the three variables being printed.

Global variables, local variables, and non local variables

Let's analyze the following code snippet:

```
def local_variable():
    """Local variable test"""
    message = "If you do what you need, you're surviving. If you do what you want, you're living."
    return message

>>> local_variable()

'If you do what you need, you're surviving. If you do what you want, you're living.'
```

This is a simple function that returns a message. Inside of the function local_variable is a variable that gets returned. This is a **local variable** as it's created inside of the function. Since its local, it can only be accessed in a limited fashion – therefore, if the variable is printed without an instance of the object then an error will occur. An example of it is listed below:

```
>>> print(message)
Traceback (most recent call last):
  File "<stdin>", line 1, in <module>
NameError: name 'message' is not defined
```

The variable can only be accessed is through an instance of the object. For example, if a message variable was created outside of the function then this is what the result would be:

```
>>> message = "Hi"
>>> local_variable()
"If you do what you need, you're surviving. If you
do what you want, you're living."
>>> message
'Hi'
```

There's another way in which local variables inside functions can be accessed without instances. The way to do this is to make the function *global*.

```
def local_variable():
    """Local variable test"""
    global message
    message = "If you do what you need, you're surviving. If
you do what you want, you're living."
    return message
```

```
>>> local_variable()
"If you do what you need, you're surviving. If you do
what you want, you're living."
>>> print(message)
If you do what you need, you're surviving. If you do
what you want, you're living.
```

What happened here is that the variable inside of the method was made a global variable via the **global** keyword. This means that the variable can now be accessed outside of the method. The following coding snippet helps explain this:

```
def local_variable():
    """Local variable test"""
    global message
    print(message)
```

```
    message = "If you do what you need, you're surviving. If
you do what you want, you're living."
    return message

>>> message = "Jambo"
>>> local_variable()
Jambo
"If you do what you need, you're surviving. If you do
what you want, you're living."

>>> print(message)
If you do what you need, you're surviving. If you do
what you want, you're living.
```

The code contains a function that has a variable that's been made global via the global keyword. The important thing is to look at the statements outside the function. What happens is that a variable is created named message which stores the string *Jambo*. Then, the local variable is called which prints Jambo in the method since it hasn't been assigned yet in the function – since message is a global variable as it's located outside of the function its value is printed. However, all subsequent method calls stores the message that's been assigned in the method as global variables which takes precedence over the ones that's located outside of it. A **nonlocal** variable is one in which you can assign values to a variable in an outer but non-global scope as shown below:

```
def outer():
    """this is the outer function."""
    x = "Hello"
    def inner():
        """ This is the inner function."""
    y = 10
    print("inner", y)
```

```
    print("outer", x)
outer()
```

```
>>> outer()
inner 10
outer Hello
```

The above shows a function named outer() that contains a local variable x, and an inner function named inner(). The inner function contains a local variable of the name y. When outer() is called the results of 10 and Hello are returned. However, what if we do the following?

```
def outer():
    """this is the outer function"""
    y = "Hello"
def inner():
    """ This is the inner function"""
    nonlocal y
    y = 10
    print("inner", y)
inner()
print("outer", y)
```

What happens is that the following is printed:

```
>>> outer()
inner 10
outer 10
```

The sole difference is the addition of the **nonlocal** keyword. A nonlocal variable is similar to a global variable except that nonlocal variables are used for variables in outer function scope, and the latter is used within a global scope.

Therefore, in this example the value of nonlocal y is applied to all statements of y within the functions. In other words, if the nonlocal y statement was removed then the following would be printed:

```
inner 10
outer Hello
```

Instance, Static, and Class Methods

We have mostly used instance methods up to this point which is when a method is binded to an object of a class. An example of an instance method is listed below:

```
import math
class Hexagon:
    """computes area of Hexagon. A = (3*sqrt(3)/2) *
side**2 """
    def area(self, side):
        return (3 * (math.sqrt(3)/2) * side **
2).__round__(2)

>>> hex = Hexagon()
>>> area = hex.area(9)
>>> area
210.44
Output: 210.44
```

There's a class named Hexagon which contains the method. An instance of the class is created, and then the method is called by using the object. Instance methods are a common part of OOP but they're many more types of methods that a person can use such as static ones. A **static method** is a type of method that has no knowledge about the class or instance it was called on. You use what's known as a decorator to make a static method.

```
import math
class Hexagon:
    @staticmethod
    def area(side):
        return (3 * (math.sqrt(3) / 2) * side **
2).__round__(2) # round to two digits

Hexagon.area(20)# 1039.23
hex = Hexagon()
area = hex.area(20) #584.57
```

The @staticmethod thingy above the method name is a
decorator that lets the compiler know that this is a static
method. With a static method you don't have to create an
instance of the class in order to call the function even
though you could if you want to. You could just use the class
name followed by a dot and then the method name. Lastly,
the self parameter is not needed. The last type of method
I'll introduce is the **class method**. To create a class method
you would also use a decorator like that with the static
method called the @classmethod. Here's the previous static
method translated to a class method:

```
import math
class Hexagon:
    @classmethod
    def area(cls, side):
        return (3 * (math.sqrt(3) / 2) * side **
2).__round__(2)

>>> Hexagon.area(20)
1039.23
>>> hex = Hexagon()
>>> area = hex.area(15)
```

```
>>> area
584.57
```

Note, `cls` is the first argument for class methods, like how `self` is the first argument for instance methods. With a class method it receives the class as an implicit first argument, just like an instance method would receive the instance. A static method does not receive no implicit first argument.

Chapter III optimized

In this chapter you were given a gentle intro to the coding paradigm known as **object oriented programming** (OOP). To help gain a better understanding of OOP let's turn to a subject we're all familiar with... people! Every human on this beautiful planet is of type homo sapien. Therefore, if we wanted to create a Python program to model the characteristics of a human then we could use Human as the class and then add data that's consistent among humans such as name, gender, birthplace, age, height, weight, and occupation. These can all be fields of the class, or variables that's declared directly in the class.

A class is just a generic template that's used to build objects, and from a single class many objects can be created. Kind of like how a person can have a complex genealogy, the same thing can be said for software that relies on the OOP approach. For example, a class could be extended and therefore have many sub classes or descendants which inherits attributes of the parent class; this is known as inheritance.

Python is a multi-paradigm language with OOP being the nucleus as everything in Python is an object. Even with this knowledge it's important to know that Python doesn't enforce developers to write OOP code all of the time compared to languages like Java, and Python also gives developers the flexibility to mix multiple coding paradigms into their code if they choose.

Chapter III Coding Challenge

They're many types of polygons, so many in fact that they can probably be explored in a separate book. To gain a better understanding of Polygons we're going to model them in Python. Create a class named Polygon that contains the attributes for a polygon such as:

- number of sides
- area
- perimeter

Methods should be created that returns the values of each of these data. Therefore, the Polygon class should have three methods:

- number_of_sides()
- area()
- perimeter()

Then, create several sub classes that are concrete implementations of specific polygons. The polygons we're going to model are:

- Triangle
- Rhombus
- Pentagon
- Hexagon
- Heptagon
- Octagon
- Nonagon
- Decagon

Each of these polygons should represent a class of a corresponding name that provides a concrete implementation. Here are the details and formulas for the following polygons that should be translated into

Python code.

Triangle: A 3-sided polygon:

$$A = \frac{1}{2}bh$$

Heron's formula

$$S = \frac{A+B+C}{2}$$

$$Area = \sqrt{S(S-A)(S-B)(S-C)}$$

In Heron's formula, S represents *semiperimeter*. This value is needed to plug into the second Area formula.

Rhombus: A 4-sided polygon:

$$A = \frac{pq}{2}$$

$$P = 4a$$

Pentagon: A 5-sided polygon.

The following formula is the area for a regular pentagon:

$$A = \frac{1}{4}\sqrt{5(5+2\sqrt{5})a^2}$$

$$P = 5a$$

Hexagon: A 6-sided polygon.

The following formula is for a regular hexagon.

$$A = \frac{3\sqrt{3}}{2}a^2$$

$$P = 6\,a$$

Heptagon: A 7-sided polygon.

The following formula is for a regular heptagon.

$$A = \frac{7}{4} a^2 \cot\left(\frac{180°}{7}\right)$$

$$P = 7\,a$$

Octagon: A eight sided polygon.

The following formula is for a regular octagon.

$$A = 2\left(1 + \sqrt{2}\right) a^2$$

$$P = 8\,a$$

Nonagon: A nine sided polygon.

The following formula is for a regular nonagon.

$$A = \frac{9}{4} a^2 \cot\left(\frac{180°}{9}\right)$$

$$P = 9\,a$$

Decagon: A ten sided polygon.

The following formula is for a regular decagon.

$$A = \frac{5}{2} a^2 \sqrt{5 + 2\sqrt{5}}$$

$$P = 10\,a$$

Here are some test cases of what the output should look like when certain methods are ran:

```
tri = Triangle()
```

```
print("Triangle Area:", tri.area(5, 10))
print("Herons formula:", tri.herons_formula(5, 4, 3))
print("Perimeter:", tri.perimeter(20, 71, 90))
print("-----------------")


rho = Rhombus()
print("Rhombus Area:", rho.area(12.3, 83.9))
print("Perimeter:", rho.perimeter(5))
print("-----------------")


pent = Pentagon()
print("Pentagon Area:", pent.area(5))
print("Perimeter:", pent.perimeter(7.5))
print("-----------------")


print("Pentagon Area:", pent.area(5))
print("Perimeter:", pent.perimeter(7.5))
print("-----------------")


hex = Hexagon()
print("Hexagon Area:", hex.area(5))
print("Perimeter:", hex.perimeter(11.25))
print("-----------------")


hep = Heptagon()
print("Heptagon Area:", hep.area(10))
print("Perimeter:", hep.perimeter(8))
print("-----------------")


oct = Octagon()
print("Octagon Area:", oct.area(10))
print("Perimeter:", oct.perimeter(7))
print("-----------------")


non = Nonagon()
```

```
print("Nonagon Area:", non.area(6))
print("Perimeter", non.perimeter(5))
print("-----------------")


dec = Decagon()
print("Decagon Area:", dec.area(10))
print(dec.perimeter(11.25))
```

```
-------------------


Triangle Area: 25.0
Herons formula: 6.0
Perimeter: 181
-----------------
Rhombus Area: 515.985
Perimeter: 20
-----------------
Pentagon Area: 43.01193501472417
Perimeter: 37.5
-----------------
Hexagon Area: 64.9519052838329
Perimeter: 67.5
-----------------
Heptagon Area: 363.39124440015894
Perimeter: 56
-----------------
Octagon Area: 482.84271247461896
Perimeter: 56
-----------------
Nonagon Area: 222.5456709758244
Perimeter 45
-----------------
Decagon Area: 769.4208842938134
112.5
```

Chapter III coding solution: http://bit.ly/2PnRBos

Chapter IV: Data Science in Python

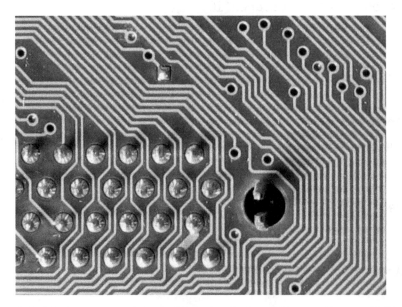

Karl-Ludwig Poggemann - labyrinthine circuit board lines — Photo - CC BY 2.0

Data science is a sizzling hot field. According to the *Harvard Business Review* in 2012, it's The Sexiest Job of the 21st Century. Even though there's no standard definition of what data science is, most definitions typically refer to it as an interdisciplinary field that encompass computer science, statistics, machine learning, research, and analytics. In order for someone to become an effective data scientist, they must be able to extract knowledge or insights from data in various forms. In this chapter you'll get an introduction to packages in Python that can be used for data science. You'll learn about the Anaconda package manager, Pandas, NumPy, and Matplotlib.

Anaconda

Anaconda is a cross-platform Python distribution for data analytics and scientific computing. With Anaconda you can install some of the most popular packages for scientific computing such as the SciPy stack (IPython, NumPy, Matplotlib, etc). Once the Anaconda installer is ran you'll have access to Pandas and the rest of the SciPy stack without needing to install anything else. Here's the Anaconda install instructions for the three major operating systems:

- Windows: http://bit.ly/2PrYky0
- OS X: http://bit.ly/2Nu834R
- Linux: http://bit.ly/2pH9aVp

To make sure that Anaconda is installed, enter the following into the terminal:

```
anaconda-navigator
```

The Anaconda Navigator should open which is a GUI for helping you start Python applications and manage the packages for your Anaconda install. It should look something like the following screenshot:

Figure 4.0: Anaconda Navigator screenshot.

Anaconda is pretty dang awesome. You have a myriad of tools at your fingertips out of the box such as JupyterLab, Jupyter Notebook, and Qt console along with the capability to easily install IDEs and tools for data visualization.

Conda tutorial for Linux

Once Anaconda is installed and you got acquainted with anaconda-navigator the next step is to get familiar with Conda which is the package manager. Anaconda is like a superset of Conda as it contains an array of packages including Conda, NumPy, SciPy, Ipython, etc. There's also something known as Miniconda which is similar to Anaconda except that it's smaller as it contains just Conda and its dependencies. Once you have Miniconda installed you can add Anaconda if you desire by running install anaconda. Anyhow, let's learn how to get acquainted with Conda – it's an open source package and environment management system for any programming language even though it's particularly popular in the python realm. Go ahead and open up the terminal and type the following:

```
conda --version
```

This should reveal the current version of Conda you're running like Conda 4.5.9

If there's a current version of Conda available you can upgrade by adding the final command into the terminal:

```
conda update conda
```

To see a list of all of the commands available in Conda type:

```
conda update —help
```

This will provide a list of all of the Conda commands along with a description of what they do. To discover what list of packages you have installed on your machine you can use the following command:

```
conda list
```

This will recursively list all of the packages that you currently have installed. To search if you have a package use the *search* command. For example, to find the package name urllib3 use the following:

```
conda search urllib3
```

If it's installed it will list all of the packages that's installed; if not, then you'll get the following error:

`PackagesNotFoundError`

Managing Environments

With the help of Conda you can create an isolated Python environment for your project – an environment is a set of packages that can be used in one or more projects. The default environment with Anaconda is the root which contains the default packages. There are two ways in which you create a Conda environment which is an environment file in YAML format (environment.yaml), or by using the create command. YAML, the acronym for "YAML Ain't Markup Language" is a human-readable data serialization language like XML or JSON. It's very popular among Python developers due to its similar use of newlines and indentation. It's important to know that Conda is compatible with pip, so if a package is not available in Anaconda then you can install them with the following:

`pip install`

The following shows an environment that will install python 3.6, python-pip, and requests. Go ahead and open up a text editor and then copy and paste the following contents into that file:

```
name: sandbox
channels:
- conda-forge
dependencies:
- python=3.6
- pip
```

```
- pip:
- requests
```

Save the file as environment.yaml, open up the terminal, change into the directory where the yaml file is located, and then run the following command:

```
conda env create -f environment.yml
```

This should start the YAML installation process. Once done you can activate the environment on Linux or OS X by using the following command:

```
source activate sandbox
```

On a windows operated machine you can simply do:

```
activate playenv
```

You can deactivate the environment on Linux/OS X by using:

```
source deactivate
```

On Windows use:

```
deactivate
```

You can also create a virtual environment by specifying the name, channel, and list of packages within the terminal. For example, to create an environment named test that uses python3.6 to install spacy, doit, dulwich, and eight, open up the terminal and enter the following command:

```
conda create -c conda-forge -n test python=3.6
spacy doit dulwich eight
```

The Different Ways to Install Panda

That's the basics of Anaconda. You can keep up to date with the updates and read through their online documentation for a more thorough understanding of it: http://docs.anaconda.com/anaconda

Pandas

This section will introduce you to the *pandas* library which is a popular tool for conducting data analysis with Python. **Pandas** can be described as an open source Python library for data manipulation and analysis. The name pandas is not inspired by the fluffy bear who feasts on bamboo, but instead "panel data" which is an econometrics term.

The Different Ways to Install Panda

If you followed the instructions at the beginning of this chapter and installed Anaconda, then you should be able to import panda directly and be up and running. Refer to the first section of this chapter for installation details with Anaconda as installing pandas with a prepackaged solution is probably the most straightforward option. Below are more options on how to install panda:

Installation from PyPI

The pandas library can be installed by PyPI using the

following command:

```
pip install pandas
```

Installation from Linux

If you're working on a Linux distro and refuse to use any prepackaged solution then you can install from the terminal:
- Debian and Ubuntu:
  - `sudo apt-get install python-pandas`
- Fedora:
  - zypper in python-pandas

You can also install it from conda by using the following command line:

conda install -f pandas

To check that pandas installed correctly type in the following statement:

```
import pandas
```

If no error occurs then pandas have been successfully installed.

Panda Data Structures: Series and DataFrames

At the focal point of panda are two data structures which are Series and DataFrame. A **Series** can be described as an arrangement of one-dimensional data containing a sequence of values, and a **DataFrame** which is the main object of panda is a two-dimensional tabular data structure.

Here's an example of a Series:

```
>>> import pandas as pd
>>> s = pd.Series([1,5,10,15,20])
>>> s

0     1
1     5
2     10
3     15
4     20
dtype: int64
```

The pandas library is imported as pd. Then, a Series is created, which is a one-dimensional labeled array that can hold a mixture of data types such as integer, floating points, strings, tuples, etc. As you can see, Series are indexed like lists or tuples. You can access the values by using the subscript notation as shown below:

```
>>> s[1]
5
>>> s[3]
15
>>> s[-1]
ERROR!
```

However, there's something that you should know which is you cannot use negative indexes in Series. Therefore, the following will create an error: s[-1]

A series has a host of attributes and methods that you can use to access data and manipulate it as shown in the following code fragment:

```
>>> s.size
5
>>> s.values
array([ 1,  5, 10, 15, 20])
>>> s.ndim
1
>>> s.imag
array([0, 0, 0, 0, 0])
>>> s.is_unique
True
>>> s.tail(2)
3    15
4    20
>>> s.to_csv('/home/usr/Desktop/Series.csv')
>>>
array([ 1,  5, 10, 15, 20])
```

DataFrame

The primary pandas data structure is known as the **DataFrame**. A DataFrame is as a rectangular table of two-dimensional, mutable, potentially heterogeneous data. It has labeled axes aka rows/columns, and according to the pandas docs, it can be thought of as a dict-like container for Series objects. A distinguishing difference between Series and DataFrames is that a Series is a data structure for a single column of data while DataFrames caters to the possibility that you'll be working with more than one column.

Here's the class signature for pandas.DataFrame:

```
class pandas.DataFrame(data=None, index=None,
columns=None, dtype=None, copy=False)
```

The DataFrame accepts five parameters but only one is mandatory. Therefore, a simple DataFrame could be constructed as follows:

```
>>> import pandas
>>> data = pandas.DataFrame([0,1,2,3])
>>> data
   0
0  0
1  1
2  2
3  3
```

There are many ways in which you can construct a DataFrame. Let's use *numpy* which you'll learn more about in the next section to create one.

```
>>> import pandas as pd
>>> import numpy as np
>>> pd.DataFrame(np.random.randint(100,
size=(4,4)),columns=['a','b','c','d'])

    a   b   c   d
0  89  53  88  40
1  16  28  90  66
2  35  59  83  24
3  20  60  73  12
```

The above creates a 4 x 4 DataFrame that randomly selects numbers in the range of 0-100. The following snippet creates a DataFrame that compares the temperatures of Los Angeles and San Francisco from 2012-2017:

```
>>> import pandas as pd
>>> dates = pd.date_range('04-03-2012','04-08-2012')
>>> cities = pd.DataFrame({'Los Angeles':
[75,72,64,72,82,79], 'San Francisco':
[63,57,54,57,64,63]}, dates)
>>> cities

            Los Angeles  San Francisco
2012-04-03           75  63
2012-04-04           72  57
2012-04-05           64  54
2012-04-06           72  57
2012-04-07           82  64
2012-04-08           79  63
```

In the above program the pandas library is imported as pd. The date_range() function is used in order to create a range of dates to analyze. Then, a DataFrame is created from two dictionaries, one to represent the city of Los Angeles during those time periods, and another to represent the city of San Fran. We can use various operators and functions to manipulate the data and generate insights from it. For example, to get the difference in temperature for each day do the following:

```
>>> cities['Los Angeles'] - cities['San Francisco']
2012-04-03    12
2012-04-04    15
2012-04-05    10
2012-04-06    15
2012-04-07    18
2012-04-08    16
Freq: D, dtype: int64
```

To return the cities back in a specific order do the following:

```
>>> cities[['San Francisco', 'Los Angeles']]

             San Francisco  Los Angeles
2012-04-03              63  75
2012-04-04              57  72
2012-04-05              54  64
2012-04-06              57  72
2012-04-07              64  82
2012-04-08              63  79
```

And, if you want to access just the temperatures of Los Angeles do the following:

```
>>> cities['Los Angeles']

2012-04-03    75
2012-04-04    72
2012-04-05    64
2012-04-06    72
2012-04-07    82
2012-04-08    79
Freq: D, Name: Los Angeles, dtype: int64
```

If you want to find the average use the mean() function of pandas.DataFrame.

```
>>> cities.mean()
Los Angeles      74.000000
San Francisco    59.666667
dtype: float64
```

Loading data from csv files into a DataFrame

Pandas makes it easy for coders to load data from the web and various sources. There's a built in function called `read_csv()` that allows you to read in data from csv files. An example of a csv file that we're going to import is a list of some of the most populous us cities from 2014-2018. You can download the **us_cities.csv** data set from github here: http://bit.ly/2A133kV

Below is the code to read the contents into a DataFrame:

```
>>> import pandas as pd
>>> file = pd.read_csv('us_cities.csv')
>>> file
   Year     New York Los Angeles   ... San Diego Dallas  San Jose
0  2018  8580015.000     4030668 ...   1438060 1359133 1030796
1  2017  8622698.000     3999759 ...   1419516 1341075 1035317
2  2016  8669000.000     4114000 ...   1436000 1427000 1049000
3  2015     8550.405     3971883 ...   1394928 1300092 1026908
4  2014  8491079.000     3947476 ...   1381069 1281047 1015785

[5 rows x 11 columns]
```

NumPy

NumPy is a library for Python that includes support for computing large multidimensional arrays and matrices. Before NumPy there was Numeric, developed by Jim Hugginn in 1995, and Numarray which was a reimplementation of Numeric– both of these packages were used for computing arrays. Travis Oliphant came along and decided to consolidate these packages which spawned the birth of NumPy in 2006 – Oliphant is also the founder of the startup Anaconda which is the package manager discussed in the first section of this book.

NumPy is now considered the library of choice for scientific computing, and has a suite of functions for performing high-level calculations on arrays in an efficient manner. At the time of publication NumPy is open sourced under the BSD license. It has a large and active community making it a smart investment in your technical career. Before we get started with NumPy we have to make sure that it's configured and running on our computer.

NumPy Installation

If you have Python installed then you may already have NumPy as it's often included in many Python distributions. To test that it's installed open up a terminal, type *python3*, and then enter the following command:

```
>>> import numpy as np
```

If nothing happens then that's great as NumPy is installed, if not then bummers, you have some work to do. There are a plethora of ways in which you can install NumPy:

- **Windows**: I'll recommend using *Conda* as a quick and easy way to get NumPy installed on your machine:
  - conda install numpy
- **Linux**: You can install NumPy through the terminal:
  - Linux(Ubuntu and Debian): sudo apt-get install python-numpy
  - Fedora: sudo yum install numpy scipy
- **OS X**: Go to the terminal and type:
  - pip3 install numpy

Once NumPy is installed, test that everything works properly by opening up the terminal and typing the following:

```
>>> import numpy as np
```

If no errors occur then NumPy is officially installed.

Learning The Lingo of NumPy

The main data structure in the NumPy library is the **ndarray** which is short for N-dimensional array. An ndarray is a multidimensional homogeneous array – homogeneous is simply a fancy word meaning that they're all of the same type. The ndarray data type utilizes another type known as **dtype** (data type) as its underlying data structure. The dtype, describes how bytes of data in the fixed-size block of memory should be interpreted; i.e., integer, float, etc. A dimension in a ndarray is defined by its **axes**. The summation of axes in a ndarray is it's **rank**, and the number of dimensions and items in an array defines its **shape –** the shape is a tuple of positive integers. Dimensions, axes, and shape are all interrelated concepts.

Like lists, the ndarray has built in attributes that allows you to obtain various information of the object such as *ndim* for getting the axes, *size* for the length, or *shape* to get its well, shape. There's also a list of functions for ndarrays that'll allow you to perform various computations on it. The first one we'll learn is the array() function which allows you to create a ndarray.

The focal point of NumPy – The ndarray

There are many paths we can take to create a ndarray, so let's start with the easiest one. A simple way to create an ndarray is to call the array() function and then pass a *list* into it as shown below:

```
>>> import numpy as np
>>> arr = np.array([1,2,3])
>>> arr
array([1, 2, 3])
```

As you can see from the code snippet the numpy library was imported as np, and then the array() function was called on np and then saved in a variable named arr. Let's explore some of the built in attributes and functions of the ndarrays:

```
>>> arr.size
3
>>> arr.shape
(3,)
>>> arr.dtype
dtype('int64')
>>> arr.ndim
1
```

The size of the array is 3 because that's how many elements it contains. The shape is the size of each dimension, and the result is a list of positive tuples; this is exactly what we got, which is the single element of 3. It's important to note that the result is (3,) instead of (1,3) because it's a 1D array. For example, here's the conventions for shape under the following circumstances:

- 1D array – tuple with 1 element – (i,)
- 2D array – tuple with 2 elements – (i,j)
- 3D array – tuple with 3 elements – (i,j,k)

Below are more functions that can be applied to a ndarray:

```
>>> arr.reshape(3,1)
array([[1],
       [2],
       [3]])
>>> arr.sum()
6
>>> arr.fill(100)
>>> arr
array([100, 100, 100])
>>> arr.put(0,250)
>>> arr
array([250, 100, 100])
```

The reshape() function changes the shape of the ndarray so in this case it's given three rows and one column, the sum() function adds up all of the elements of the array, the fill() function fills the elements with a certain value, and the put() function places a value at a certain position in the array. Arrays can be created to have many dimensions. After all, NumPy is described as the library for creating large and multidimensional arrays. The following shows how to create a 2D ndarray:

```
>>> arr = np.array([[1,2,3],[4,5,6]])
>>> arr
array([[1, 2, 3],
       [4, 5, 6]])
>>> arr.ndim
2
```

As you can see the number of square brackets you pad onto the array represents the number of dimensions it has. Therefore, what do you think will be the dimension of the following ndarray?

```
>>> arr = np.array([[[1,2,3],[4,5,6],[7,8,9]]])
```

The answer is three. What about the shape? The answer is...

```
>>> arr.shape
(1, 3, 3)
```

Manipulating ndarrays

The array() function is not the only function we can use to create an array. In most arrays, the elements are initially unknown, but the size of it is known. Therefore, there are several functions for creating arrays with initial placeholder content. This helps combat the need to grow an array which is an expensive operation and can cause performance issues. There are three functions you should be aware of when it comes to creating placeholder content for arrays. The first one is zeros() which pad the array with zeroes, the second is ones() which pads it with 1s, and the last one is empty() which randomly fills the array with numbers.

```
>>> np.zeros([3,4])
array([[0., 0., 0., 0.],
       [0., 0., 0., 0.],
       [0., 0., 0., 0.]])
```

```
>>> np.ones([7,7])
array([[1., 1., 1., 1., 1., 1., 1.],
       [1., 1., 1., 1., 1., 1., 1.],
       [1., 1., 1., 1., 1., 1., 1.],
       [1., 1., 1., 1., 1., 1., 1.],
       [1., 1., 1., 1., 1., 1., 1.],
       [1., 1., 1., 1., 1., 1., 1.],
```

```
    [1., 1., 1., 1., 1., 1., 1.]])

>>> np.empty([3,5])
array([[7.104960e-318, 7.013341e-318, 7.094783e-318,
6.891860e-318,
        6.823387e-318],
       [6.715009e-318, 6.625791e-318, 7.050317e-318,
6.423308e-318,
        6.329213e-318],
       [5.092809e-318, 5.115146e-318, 5.182749e-318,
5.073600e-318,
        5.018645e-318]])
```

You can also pass in more than two parameters when creating these arrays. For example, look at the following code snippet:

```
>>> np.zeros([3, 2,3])
array([[[0., 0., 0.],
        [0., 0., 0.]],

       [[0., 0., 0.],
        [0., 0., 0.]],

       [[0., 0., 0.],
        [0., 0., 0.]]])
```

What happens is the last two parameters represent the rows and columns, while the first parameter states how many of these arrays to create. You can also create arrays using the arange() function to return an even sequence of values. Here's how the function signature looks:

```
numpy.arange([start,]stop,[step,]dtype=None)
```

This function has four parameters, only one which is

mandatory. This function works in a similar manner to the range() function that's part of the Python core. For example, to print the numbers from 0...19 you'll do the following:

```
>>> np.arange(20)
array([ 0,  1,  2,  3,  4,  5,  6,  7,  8,  9, 10,
11, 12, 13, 14, 15, 16, 17, 18, 19])
```

You can also specify the start and stop intervals as indicated below:

```
>>> np.arange(5,10)
array([5, 6, 7, 8, 9])
```

Last, you can specify the step that the function takes by adding one more parameter as indicated below:

```
>>> np.arange(1,51, 4)
array([ 1,  5,  9, 13, 17, 21, 25, 29, 33, 37, 41,
45, 49])
```

You don't have to use an integer as the step, you can also use a floating point as indicated below:

```
>>> np.arange(1,5,.5)
array([1. , 1.5, 2. , 2.5, 3. , 3.5, 4. , 4.5])
```

The above example allows us to create arrays, but the problem is that they're all single dimensional arrays. What if we want to create multi dimensional arrays? We can, we could chain the arange() and reshape() functions. Let's assume that we had an array of size 10. What are the possible ways that we could reshape the array? Well, the ndarray should have an even amount of elements, so one way we can do this is 10/2 or 2 x 5, or 10/5 or 5 x 2. This is how it looks programmatically:

```
>>> arr_1 = np.arange(1,11).reshape(2,5)
>>> arr_2 = arr = np.arange(1,11).reshape(5,2)
>>> arr_1
array([[ 1,  2,  3,  4,  5],
       [ 6,  7,  8,  9, 10]])
>>> arr_2
array([[ 1,  2],
       [ 3,  4],
       [ 5,  6],
       [ 7,  8],
       [ 9, 10]])
```

There's two more functions that we can use to create ndarrays in Python which are linspace() and random(). The linspace() function is similar to the arange() function with the inclusion of an optional third element which allows you to specify the distance between one element and the next. For example, check out the following:

```
>>> np.linspace(1,10,6)
array([ 1. ,  2.8,  4.6,  6.4,  8.2, 10. ])
```

An easy way to remember the functionality of this function is that the third parameter denotes how many elements will be in the result. If you include a floating point number as the third parameter then it will be converted to an integer and rounded down as shown below:

```
>>> np.linspace(1,10,7.29)
array([ 1. ,  2.5,  4. ,  5.5,  7. ,  8.5, 10. ])
```

The other option you can use is random() which as the name indicates allows you to randomly construct your array:

```
>>> np.random.random(10)
array([0.16226991, 0.71935471, 0.98328975, 0.77536231,
0.62237784,
       0.85478772, 0.97055716, 0.10275402, 0.21966469,
0.78476643])
```

If you want to create a multidimensional array then you can pass the size of the array as shown in the snippet below:

```
>>> np.random.random((5,3))
array([[0.3219124 , 0.67128438, 0.59852726],
       [0.51625041, 0.53455118, 0.3961402 ],
       [0.72559714, 0.98484787, 0.5868404 ],
       [0.0946395 , 0.84296275, 0.18282056],
       [0.04034345, 0.12749643, 0.33662377]])
```

Let's take a look at some mathematical operations that we can perform on ndarrays.

```
>>> c = np.array([[1,2,3],[4,5,6], [7,8,9]])
>>> d = np.array([[1,1,1],[1,1,1],[1,1,1]])
>>> c + d
array([[ 2,  3,  4],
       [ 5,  6,  7],
       [ 8,  9, 10]])
```

As you can see in the above example that two arrays are created and summed together. It doesn't matter if the arrays are not the same dimensions as they can still be summed together. For example, look at the following code snippet:

```
>>> e = np.array([1,2,3])
>>> d + e
```

```
array([[2, 3, 4],
       [2, 3, 4],
       [2, 3, 4]])
```

The same thing can be said for subtraction as shown below:

```
>>> d - e
array([[ 0, -1, -2],
       [ 0, -1, -2],
       [ 0, -1, -2]])

>>> d
array([[1., 1., 1.],
       [1., 1., 1.],
       [1., 1., 1.]])

>>> e
array([1, 2, 3])

>>> f = d * e
>>> f
array([[1., 2., 3.],
       [1., 2., 3.],
       [1., 2., 3.]])
```

Let's turn our attention to linear algebra which is found in the numpy.linalg package. It includes functionality for matrix/vector products, decomposition, eigenvalues, and solving equations. Let's explore some of this below...
The dot product:

```
>>> a = np.array([1,2,3])
>>> b = np.array([66, 273, 9])
```

```
>>> np.dot(a,b)
639
```

Matrix production:

```
>>> a = np.array([1,2,3])
>>> b = np.array([4,5,6])
>>> np.matmul(a,b)
32
```

Matrix transposition:

```
>>> a = np.array([[1,2,3], [4,5,6]])
>>> a.transpose()
array([[1, 4],
       [2, 5],
       [3, 6]])
```

There are many functions that can be applied to ndarrays. They can be classified as universal functions (ufunc), and aggregate functions. An **ufunc** is a type of function that operates on ndarrays in an element-by-element fashion. Examples of such functions are sin(), cos(), log(), and sqrt() shown below:

```
>>> arr = np.arange(1,6)
>>> arr
array([1, 2, 3, 4, 5])
>>> np.sin(arr)
array([ 0.84147098,  0.90929743,  0.14112001, -0.7568025
, -0.95892427])
>>> np.cos(arr)
array([ 0.54030231, -0.41614684, -0.9899925 , -
0.65364362,  0.28366219])
```

```
>>> np.log(arr)
array([0., 0.69314718, 1.09861229, 1.38629436,
1.60943791])
>>> np.sqrt(arr)
array([1., 1.41421356, 1.73205081, 2.        ,
2.23606798])
```

An **aggregate** function is one that sums the values of the array such as `sum()`, `min()`, `max()`, and `mean()` as shown below:

```
>>> arr = np.random.random(5)
>>> arr
array([0.82805426, 0.39515967, 0.88651131, 0.63475258,
0.24256676])
>>> arr.sum()
2.987044585482207
>>> arr.min()
0.24256675908481473
>>> arr.max()
0.8865113126777416
>>> arr.mean()
0.5974089170964414
>>> arr.max()**2 - arr.min()/arr.mean()
0.37987094080999884
```

Accessing Elements

Once you've learned how to create ndarrays and execute various operations on them the next step is to learn how to manipulate them. Like lists, you can index, slice, and iterate over ndarrays. For example, check out the following ndarray:

```
>>> x = np.array([3728, 883, -299, 0, 983, 3, 65,
.00283])
```

How would you access element -299? One solution is as below:

```
>>> x[2]
-299.0
```

How about the last element? You could use x[7], or you could use negative subscript notation and do the following: x[-1]

So, accessing elements for ndarrays are identical to accessing elements for lists. What about when you have multidimensional arrays? An example of how to do this is listed below...

```
>>> y = [[4,9,10],[53,87,10],[7,62,982]]
>>> np.reshape(y,[3,3])
array([[  4,   9,  10],
       [ 53,  87,  10],
       [  7,  62, 982]])
```

How would you access 9? The correct way to doing this is listed below:

```
>>> y[0][1]
9
```

This is a two dimensional ndarray so therefore you'll need to access the row and then columns. In this example there are three rows, with the first one starting at 0, and the last one starting at 3. The columns follow the same convention as it starts at 0 and ends at 2. You'll access 982 by using y[2][2].

Slicing

Like lists, you can slice an ndarray by using the colon [:] syntax, but unlike lists the arrays created by slicing are not copies. Let's look at the following array as an example:

>>> a = np.arange(1,11)

The output is:

>>> a
array([1, 2, 3, 4, 5, 6, 7, 8, 9, 10])

The first element corresponds to index 0, and the last element corresponds to the size of the list which in this case is 10. Therefore, if you want to slice the elements from 6 through 10, one solution is to use the following:

>>> a[5:10]
array([6, 7, 8, 9, 10])

It's important to know that the starting element is inclusive, and the ending one is exclusive. For example, when you type a[2:5] you'll get 3, 4, and 5 as the sixth element or 6 is excluded. If you want to start at an index and iterate all the way to the end then you can use the following notation a[start:]. For example, if we start at 5 and want to create a slice all the way to the end then we could do this:

>>> a[4:]
array([5, 6, 7, 8, 9, 10])

This way we don't have to keep track of where we have to end, we'll just know where we have to start. Also, you can specify the step in each sequence with a third parameter as shown in the following examples:

```
>>> a[1::2]
array([ 2, 4, 6, 8, 10])
>>> a[0::2]
array([1, 3, 5, 7, 9])
```

In a[1::2] the array is sliced at its index 1 which is 2, and then it increments by 2 every subsequent step until it reaches the end of the list or at 10. In a[0::2], the list starts at 1, and increments by 2 all the way up to the end of the list which in this case is 9. Let's look at some more examples that deal with slicing ndarrays:

```
>>> a[::2]
array([1, 3, 5, 7, 9])
>>> a[::3]
array([ 1, 4, 7, 10])
>>> a[::4]
array([1, 5, 9])
```

In the above examples the slicing starts at index 0, and then goes all the way to the end of the list using the third argument as the increment. With this syntax in mind you can creatively slice lists in a myriad of formats as shown below:

```
>>> a[5::2]
array([ 6,  8, 10])
>>> a[:5:]
array([1, 2, 3, 4, 5])
>>> a[:5:2]
array([1, 3, 5])
```

Now, let's turn our focus towards slicing two dimensional

arrays. The same concept applies, but the difference is that we have to slice the rows and columns separately. Let's take a look at the following example:

```
>>> i = np.arange(1,41).reshape(10,4)
>>> i
array([[ 1,  2,  3,  4],
       [ 5,  6,  7,  8],
       [ 9, 10, 11, 12],
       [13, 14, 15, 16],
       [17, 18, 19, 20],
       [21, 22, 23, 24],
       [25, 26, 27, 28],
       [29, 30, 31, 32],
       [33, 34, 35, 36],
       [37, 38, 39, 40]])
```

In the above example we have an ndarray with ten rows and four columns. What would the syntax be if we wanted to slice just the first row? The answer would look like the following:

```
>>> i[:1]
array([[1, 2, 3, 4]])
```

Therefore, with this in mind if you wanted to slice through every other row in the matrix you would use the following syntax:

```
>>> i[::2]
array([[ 1,  2,  3,  4],
       [ 9, 10, 11, 12],
       [17, 18, 19, 20],
       [25, 26, 27, 28],
       [33, 34, 35, 36]])
```

Ok, things are making sense, but how do one go about slicing columns? Well, columns like rows have indexes, and they also start at 0. Therefore, to access all of the elements of the first column of i, the syntax would be as follows:

```
>>> i[:,0]
array([ 1,  5,  9, 13, 17, 21, 25, 29, 33, 37])
```

If you wanted to access just the first five elements of the first column, the syntax would instead be:

```
>>> i[0:5,0]
array([ 1,  5,  9, 13, 17])
```

Iterating over ndarrays

Let's assume that we have an ndarray defined as follows:

```
>>> x = np.arange(1,11).reshape(5,2)
>>> x
array([[ 1,  2],
       [ 3,  4],
       [ 5,  6],
       [ 7,  8],
       [ 9, 10]])
```

How can we iterate over the array so that we can access the elements when needed? Well, you could consider using the built in constructs of the array such as a for loop. When you use a normal for loop you can traverse all of the rows of a ndarray as shown below:

```
for elements in x:
```

```
print(elements)
```
...
```
[1 2]
[3 4]
[5 6]
[7 8]
[9 10]
```

If you need to access the elements of the first column you can use a for loop as follow:

```
for elements in x:
    print(elements[0])
```
...
```
1
3
5
7
9
```

To get the elements of the second column use the following syntax:

```
for elements in x:
    print(elements[1])
```
...
```
2
4
6
8
10
```

You could also use the flat property to iterate over the array step-by-step.

```
for item in x.flat:
    print(item)
...
1
2
3
4
5
6
7
8
9
10
```

Below is a simple example that shows how to sum the elements of each column:

```
col_1 = 0
col_2 = 0
for elements in x:
    col_1 += elements[0]
    col_2 += elements[1]

>>> print("summation of col_1 =", col_1)
summation of col_1 = 25
>>> print("summation of col_2 =", col_2)
summation of col_2 = 30
```

NumPy provides an alternative to the for loop which is the apply_along_axis() function. This accepts three parameters and the first three arguments of the function are the aggregate function, the axis, and the array – if the axis is equal to 0 (default) the function is applied on the columns, and if it's equal to one then it's applied on the rows as shown below:

```
>>> np.apply_along_axis(lambda x: (x*3)/
2,axis=1,arr=x)
array([[ 1.5,  3. ],
       [ 4.5,  6. ],
       [ 7.5,  9. ],
       [10.5, 12. ],
       [13.5, 15. ]])
```

Input and output with NumPy

An important aspect of NumPy is the ability to read the data contained in a large file. This is critical, as often data scientists are working with data sets that contains a massive amount of data. The good news is that NumPy provides several functions that allow you to save the results in a binary text file, and it enables the reading of data into an array. The functions that you'll need to use for this are save() and load(). To export the array you've created you can use the save() function which accepts two arguments: the name of the array and the array you want to export. The .npy extension will be appended to the end of the file by default. Below is an example on how to use the save() function in Python.

```
>>> np.save('Data', arr)
```

When you need to retrieve the data stored in a .npy file you can use the load() function by passing the filename as the argument with the .npy extension appended to the end of the file as shown below: -1

```
>>> np.load('Data.npy')
array([[ 0,  1,  2,  3,  4,  5,  6,  7,  8,  9],
       [10, 11, 12, 13, 14, 15, 16, 17, 18, 19],
       [20, 21, 22, 23, 24, 25, 26, 27, 28, 29],
       [30, 31, 32, 33, 34, 35, 36, 37, 38, 39],
       [40, 41, 42, 43, 44, 45, 46, 47, 48, 49],
       [50, 51, 52, 53, 54, 55, 56, 57, 58, 59],
       [60, 61, 62, 63, 64, 65, 66, 67, 68, 69],
       [70, 71, 72, 73, 74, 75, 76, 77, 78, 79],
       [80, 81, 82, 83, 84, 85, 86, 87, 88, 89],
       [90, 91, 92, 93, 94, 95, 96, 97, 98, 99]
```

You can also read and save data in other formats as well. You may decide that binary format is not the best choice because you need for the files to be accessed outside NumPy. For example, a teacher may have a list of students in a class along with their grades. Let's assume that the file they have is in CSV (Comma Separated Values) format and they want to read it into NumPy as a text file and then perform calculations on it. Let's look at the CSV file contents below:

student_id	Test_1	midterm	final
1	73.8	80.26	92.39
2	50	89.2	73.2
3	92.3	86.29	95.32

You can download the file **grades.csv here**:
http://bit.ly/2CwlH6n

The function to use for this is genfromtxt(), and here's an example of it in action:

```
>>> data = np.genfromtxt('grades.csv', delimiter=',',
names=True)
>>> data
array([(1., 73.8, 80.26, 92.39), (2., 50. , 89.2 , 73.2 ),
```

```
    (3., 92.3, 86.29, 95.32)],
    dtype=[('student_id', '<f8'), ('Test_1', '<f8'),
('midterm', '<f8'), ('final', '<f8')])
```

The output displays the text in tuples, and we can access specific portions of it as shown below:

```
>>> data['student_id']
array([1., 2., 3.])
>>> data['Test_1']
array([73.8, 50. , 92.3])
>>> data['midterm']
array([80.26, 89.2 , 86.29])
```

We can use the built in functions to extract various insights from the data such as the *mean* and the *highest* and *lowest* scores.

```
>>> np.mean(data['midterm'])
85.25
>>> np.max(data['midterm'])
89.2
>>> np.min(data['midterm'])
80.26
```

Financial equations

There's a plethora of basic financial equations in NumPy that we can explore in the numpy.fv package. Let's take a look at the fv function which has the following signature:

```
numpy.fv(rate, nper, pmt, pv, when='end')
[source]
```

Further explanation of the signature will define the following parameters as follows:

- **rate**: The rate of interest as a percentage.
- **nper**: The total compounding periods. If interest happens once a year then it's yearly or *1*, and if it happens monthly then it should be *12*.
- **pmt**: Payment.
- **pv**: Present value.

All of the parameters except when can be scalars (real numbers) or arrays. Here's some examples on how to put the fv() function in action:

What's the future value after 5 years of saving $500 now, with an additional monthly savings of $50? The interest rate is 7% annual compounded monthly.

This can be accomplished using the following:

```
>>> numpy.fv(.07/12,5*12, -50, -500)
4288.457712212625
```

The negative sign represents cash flow that's going out, or money not available today. Therefore, saving $50 a month at 7% annual interest provides $4288.26 after five years.

Here's another example using the nper() function. Below is how the function signature looks:

```
numpy.nper(rate, pmt, pv, fv=0, when='end')[source]
```

- **rate**: rate of interest
- **pmt**: payment

- **pv:** present value
- **fv:** future value (set to 0 by default)
- **when:** when payments are due, optional

Let's say that you had to pay $250 per month on your student loans. How long would it take to pay it off if the loan amount was $25,000 and if you had 6.8% interest? Here's the code to compute this:

```
>>> import numpy as np
>>> months = np.nper(.068/12,-250,25000)
>>> months
array(147.99091097)
>>> years = months/12
>>> years
12.332575914446169
```

Matplotlib

A very important part of data science is visualization, and this is precisely what Matplotlib was designed for. Matplotlib is a plotting library for Python and its numerical mathematics extension of NumPy. To make our first graph in Python is super simple, and it only requires three steps:

Step one: Import matplotlib.
Step two: Enter the data to be plotted.
Step three: Display the data.

```
>>> from matplotlib import pyplot as plt
>>> plt.plot([5,6,7,8],[7,3,6,9])
[<matplotlib.lines.Line2D object at
0x7f654f930358>]
```

```
>>> plt.show()
```

The graph should display and look likes the following screenshot:

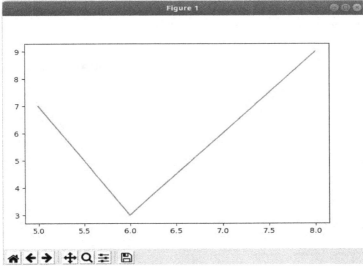

Figure 4.1: Matplotlib basic plot.

It's important to note that plt.show() must be there in order for the graph to be displayed. This is the basics of plotting your first graph using Matplotlib, but there are a lot more nifty things that you can do with this graphing library. Matplotlib features are similar to MATLAB. It supports multiple windowing environments such as GTK, Tkinter, Qt, and wxWindows as well as several non-interactive backends like pdfs.

Learning the api of Matplotlib

To see a list of all of the cool things that can be done with Matplotlib you can take a look at their documentation: http://bit.ly/2Ka9NAf

However, sometimes online docs are ironically not the best way to learn an api. It can be written in an overly technical manner or perhaps the examples could be lacking. Lets learn how to build awesome plots in Python using Matplotlib piece-by-piece. Let's build on top of the example that we had before. Let's say that you want to add a label to the x-axis. The code for doing this is listed below:

```
>>> plt.plot([5,6,7,8],[7,3,6,9])
[<matplotlib.lines.Line2D object at
0x7f4b772e3c50>]
>>> plt.xlabel("The X-Axis")
Text(0.5,0,'The X-Axis')
>>> plt.ylabel("The Y-Axis")
Text(0,0.5,'The Y-Axis')
>>> plt.show()
```

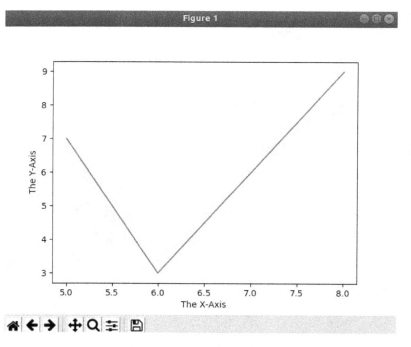

Figure 4.2: Matplotlib x-and-y axes labels.

As you can see in the updated graph the x and y axes now have labels. You can replace the text with any string you want. There's also more formatting options you can apply to the graph. For

example, you can add an axis and a title. To add a title use
plt.title() and to add an axis use plt.axis().

Let's analyze the documentation of the plot() function.
Here's how the method looks:

```
matplot.pyplotlib.plot(*args, **kwargs)
```

For a recap from chapter two, *args means that a function
enables a non-keyword argument a variable length of time,
and **kwargs allows you to process named arguments aka a
dictionary. An example on how to use the plot() function
to modify the plot is listed below:

```
>>> import matplotlib.pyplot as plt
>>> plt.plot([0,1,2,3,4],[3,6,9,12,15], "c*")
>>> plt.axis([0,10,0,20])
>>> plt.show()
```

The plot that's displayed is listed below:

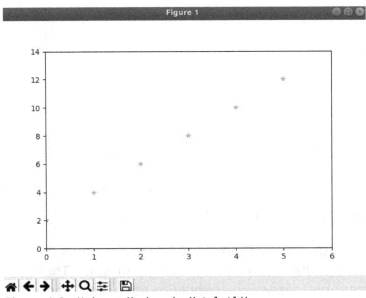

Figure 4.3: Using a Marker in Matplotlib.

As you can see from the code snippet the `plot()` function needs to be able to accept an unlimited number of arguments to adhere to the various types of plots. The number of elements in x must match the number of elements in y or less an error will occur. For example, the following will cause an issue:

```
>>> plt.plot([1,2,3],[2,4])
```

Here's what the error message say:

```
ValueError: x and y must have same first
dimension, but have shapes (3,) and (2,)
```

In order to fix this issue the length of both lists must be equal as corrected in the following coding snippet:

```
>>> plt.plot([1,2,3],[2,4,6])
```

Also, if you're confused about what "c*" stands for then just remember that these are optional parameters that control the **line style** or marker. The character 'c' represents the color *cyan*, and '*' is a star marker which means that there will be stars instead of points.

- Check out the colors in Matplotlib: http://bit.ly/2y3I0l9
- Matplotlib markers: http://bit.ly/2zYUZkH

You could alternatively format the plot by using the following syntax.

```
>>> plt.plot([0,1,2,3,4],[3,6,9,12,15],
color="cyan", marker="p")
```

Go ahead and modify the line style, color, and axis. In Matplotlib you're not limited to just lists as numeric processing can be done. All sequences are behind-the-scenes converted to NumPy arrays. Let's create a plot with NumPy as shown below:

```
>>> import numpy as np
>>> import matplotlib.pyplot as plt
>>> a = np.arange(0,10,2)
>>> plt.plot(a,a**2,'co', a,a**3,'mp', a, a**4,
'kH')
>>> plt.show()
```

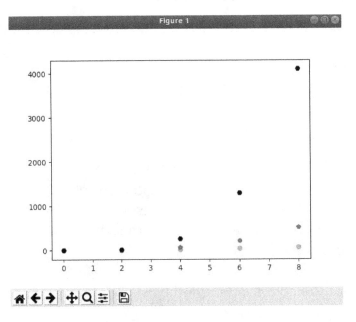

Figure 4.4: Plotting NumPy arrays.

The only new addition we added was the use of the `arange()` function of NumPy which allows us to generate a range of numbers. In the `plot()` function we're passing three x,y pairs. All of the x-pairs are identical and their corresponding y-pairs takes the x-values and scale them by 2, 3, and 4 respectively. The following code shows how to create a sinusoidal plot:

```
>>> import matplotlib.pyplot as plt
>>> import numpy as np
>>> si = np.arange(0,100)
>>> plt.plot(np.sin(si))
>>> plt.show()
```

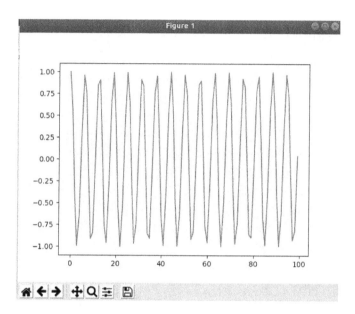

Figure 4.5: Sinusoidal waves plot.

Bar chart

There are many kinds of plots that you can create with Matplotlib such as bar, line, and pie charts. Let's get acquainted with all of them starting with the bar chart; below is its method signature:

```
matplotlib.pyplot.bar(*args, **kwargs)[source]
```

Below are the call signatures for the function:

```
bar(x, height, *, align='center', **kwargs)
bar(x, height, width, *, align='center', **kwargs)
bar(x, height, width, bottom, *, align='center',
**kwargs)
```

To view the details of the function you can check out its details here: http://bit.ly/2RA2cy8

The below code plots a simple bar chart:

```
>>> import matplotlib.pyplot as plt
>>> import numpy as np
>>> x = np.arange(1,5) # ndarray of 1-4
>>> height = x ** 2     # square elements in x
>>> plt.bar(x, height, width = 0.35, align
="center", color ="green", edgecolor="black")
>>> plt.title("Bar Graph Demo")
>>> plt.show()
```

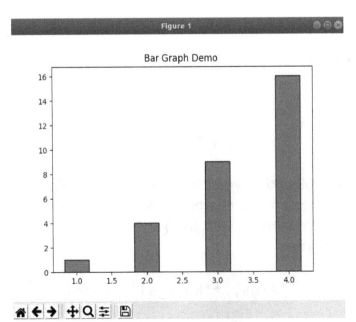

Figure 4.6: Bar Graph Demo.

Pie chart

The function signature for creating a pie chart is listed below:

```
matplotlib.pyplot.pie(x, explode=None, labels=None,
colors=None, autopct=None, pctdistance=0.6,
shadow=False, labeldistance=1.1, startangle=None,
radius=None,counterclock=True, wedgeprops=None,
textprops=None, center=(0, 0), frame=False,
rotatelabels=False, hold=None, data=None)
```

As you can see, there's a helluva arguments that you can pass into this bad boy. However, many of these parameters are initially set to None or False. The below code snippet reveals how to create a simple pie chart with Matplotlib.

```
>>> import matplotlib.pyplot as plt
>>> labels =
['HP','Samsung','Dell','Asus','Lenovo','Acer']
>>> sizes = [10,15.25,35.2,20,10.521,9.029]
colors =
['rosybrown','lightseagreen','darkorange','darkolivegree
n','mediumblue','saddlebrown']
# Plot
>>> plt.pie(sizes, labels=labels,colors=colors,
shadow=True)
#add legend
>>> plt.legend(labels)
>>> plt.show()
```

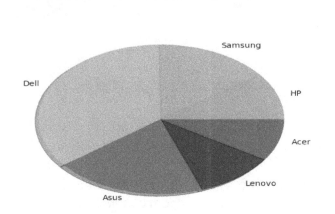

Figure 4.7: Pie Graph Demo.

Scatter plot

Below is the signature for a scatter plot in Python:

```
matplotlib.pyplot.scatter(x, y, s=None, c=None,
marker=None, cmap=None, norm=None, vmin=None, vmax=None,
alpha=None, linewidths=None, verts=None,
edgecolors=None, hold=None, data=None, **kwargs)[source]
```

Below is the code for a simple scatter plot:

```
>>> import matplotlib.pyplot as plt
>>> import numpy as np
>>> x = np.arange(1,51)
>>> y = np.random.rand(50)
>>> colors = np.random.rand(50)
>>> plt.scatter(x,y,c=colors)
>>> plt.show()
```

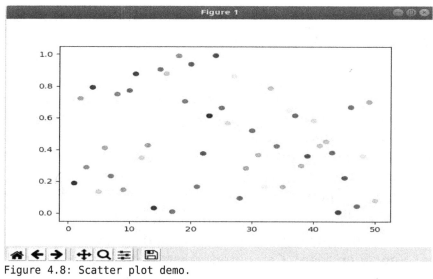

Figure 4.8: Scatter plot demo.

Chapter IV Optimized

Data science is a broad field and in this chapter you've got an appetizer of some of the popular libraries in Python for data manipulation, scientific computing, and data visualization. The Pandas library is an open source library for providing simple high performance data structures and data analysis tools for the Python programming language. NumPy is a library that adds support for large multidimensional arrays and matrices, along with a collection of high-level functions to apply on them. Compared to lists, NumPy arrays are more efficient as they require less memory, and they can also read/write items more efficiently. Matplotlib is a popular plotting library which can be used to visualize your data. Since NumPy is the numerical mathematical extension of NumPy, it's designed to work nicely with it.

Chapter IV Coding Challenge

Challenge 1: Matplotlib provides the ability to create shapes. Create a green circle with a .50 radius in Matplotlib.

Challenge 2: Plot a histogram that draws 500 random samples from a normal Gaussian distribution. The histogram should have a bin of 50, be green, and include a ylabel with the name *Probability*.

Challenge 3: Learn about 3D plotting in Matplotlib by reading the mplot3d tutorial: http://bit.ly/2zYWcZh
After reading the tutorial create a 3D line, bar, and scatter plot.

View the solutions to the chapter IV coding challenges:
http://bit.ly/2y8Kr0y

Chapter V: Data Manipulation in Python

Charles Hutchins - 4.aiff.au - CC BY 2.0

Data manipulation is the process of modifying data so that it's in a more organized state and therefore easier to handle. One example in which data manipulation is handy is web server logs. It's much easier to derive meaningful stats if it's organized nicely like which page has the most visitors or bounce rates. Python is an excellent language for data manipulation and there's some built in modules that allows you to do it out of the box. The first one we're going to play with is the CSV module – CSV files are the most common import and export format for spreadsheets and databases so it's a good idea to get familiar with them. You'll also learn how to manipulate JSON files along with using the BeautifulSoup4 library to parse HTML and XML files.

The CSV module

A **Comma Separated Values** file (CSV not CVS like the store) can be defined as a delimited file that uses commas to distinguish values. CSV files have roots dating back to the early 70s with mainframe computers, and it's still popular today being the defacto data exchange format for spreadsheets and databases. A CSV file stores tabular data in a plain text file, with each line being a data record – each record has one or more fields that's separated by commas.

The use of the comma as a file separator is where the file gets its name, even though confusingly other delimiters such as semicolons, colons, spaces, and tabs are permitted. This brings another point about CSV files which is that there's no official standardization of this data exchange format. A proposal was made in *RFC 4180* but it just talks about the standard specific handling text based fields; it don't actually talk about interpretation of the text which is application specific. Below is an example of how a CSV file may look:

Name	Role	Salary
Julius Caesar	CEO	500000
Chalemagne	CPO	200000
Joan of Arc	CFO	300000
Napoleon Bon	CMO	350000

You can download the file from github:
http://bit.ly/20IY9Ba

The good thing is that Python works hard for developers and has a module that's helpful for handling your read and write needs. This allows Python programmers to write code that reads or exports excel files without knowing the precise details of the CSV format used. In addition, programmers can define the CSV format used by other applications or create their own special purpose CSV format. The csv module which we'll discuss in detail momentarily reads and writes sequences. If coders need to read and write data from a dictionary then they can use the DictRead and DictWriter classes.

How to read in csv files

The two key functions in the csv module are reader() and writer(). Let's observe the signature for the reader() function:

```
csv.reader(csvfile, dialect='excel', **fmtparams)
```

The only required argument is csvfile; dialect and fmtparams are optional. While csvfile is typically a file it can also be an object that supports the iterator protocol and returns a string every time the next() function is called; therefore, file objects and lists are suitable. If csvfile is a file object, then it must be opened with the 'b'(binary) flag on platforms where it makes a difference. The dialect parameter which is optional can be used to define a set of parameters specific to the csvfile. Last, the fmtparams argument can be used to override the individual formatting parameters in the current dialect. The below program shows how to read in a CSV file using the reader() function.

```
import csv
with open('startup.csv', 'r') as file:
    employees = csv.reader(file)
    for row in employees:
        print(" : ".join(row))
```

```
. . .
Name : Role : Salary
Julius Caesar : CEO : 500000
Chalemagne : CPO : 200000
Joan of Arc : CFO : 300000
Napoleon Bonaparte : CMO : 350000
Frank Roose : CTO : 375000
Mao Dong : CIO : 400000
```

Here's a recap of what's happening in the above program. The csv module is imported, and then the file is opened via the open() function. The filename is passed as the first argument and 'r' indicates that the file is open for reading; there's other modes such as 'w' for writing, 'a' for appending, and 'b' for binary. You can also use r+ to indicate that the file is open for reading and writing, and r+b to open the binary file in read or write mode. If you're interested in learning more you can read about it here: http://bit.ly/2OIY9Ba

When you pass the file into the csv.reader() function, the file is read, and the for loop is used to traverse it. The join() function returns a string in which the elements has been joined by a separator or colon in this case. If you want to write a file then you'll need to use the csv.writer() function. Below is how the signature looks:

```
csv.writer(csvfile, dialect='excel', **fmtparams)
```

Also, lets discuss the with keyword. Introduced in Python 2.5, it's not mandatory, but it's a good practice to include when reading and writing files as it ensures that *clean-up* code is done such as closing resources. You can use it as a substitute for the try/except/finally blocks. In order to write a file use the writer() function whose signature is listed below:

```
csv.writer(csvfile, dialect='excel', **fmtparams)
```

Here's an example of the function in action:

```
import csv
with open('software_company.csv', 'w') as employees:
    file = csv.writer(employees)
    file.writerow(["John Q", "lead software engineer",
80000])
    file.writerow(["Anon Pikes", "software engineer",
75000])
    file.writerow(["Marky M", "business analyst", 65000])
    file.writerow(["Don L", "software tester", 65000])
    file.writerow(["Lisa Z", "web developer", 72000])
    file.writerow(["Melissa J", "marketer", 50000])
    file.writerow(["Daniel K", "human relations", 55000])
```

The csv module is imported and then the file is opened via the the open() function – remember, the with statement is used to handle any cleanup. Next, the file object, aka employees is passed to the writerow() function and then several writerow() function calls are made which write the rows to the csv table. The signature of the `writerow()` function is listed below:

```
csvwriter.writerow(row)
```

The function writes the row of the file object, formatting it in its according dialect. This function returns a list, hence the contents of the file to be written are passed as a list. The above code overwrites a file named: `software_company.csv`

Now that we have a good idea of how to read and write files using the csv module let's explore more of the things that we can do with it. Once all of the `writerow()` functions are called the file is outputted. If you were to type the above code fragment into the Python Shell you would get something like the following:

```
37
36
32
29
28
26
32
```

These numbers just represent the amount of bytes that's written to the file. Once the code successfully executes, the csv file which contains the data should be updated with the new contents. A file that's been written can therefore be read using the `reader()` function. Now that we have a general idea about how to read and write files in Python we can now explore more of the functionality of the csv module which you can analyze here: http://bit.ly/20akwQq

You can also read and write files in the csv module using the DictReader and DictWriter classes. The `DictReader()` signature is shown below:

```
class csv.DictReader(f, fieldnames=None,
restkey=None, restval=None, dialect='excel',
*args, **kwds)
```

As you can see DictReader() has quite a bit of parameters;
seven to be exact, but only the first one is required. The
above creates an object which operates like the normal
reader except that it maps the information read into a
dictionary whose keys are given by the optional fieldnames
parameter. Below is the contents of a simple file called
names.csv: http://bit.ly/2yrHdVM

The following code snippet will read in the contents of the
file using DictReader() as shown below:

```
import csv
with open('names.csv', 'r') as files:
    names = csv.DictReader(files)
    for row in names:
        print(row['first_name'], row['last_name'])
```

Here's the output of names.csv:

```
Tiffany Fitzhugh
Tanya Krotki
Rayhan Neel
Tommy Forrester
Keir Lenard
Aurelia Hsiao
oy Ellwood
Diya Graham
Sheila Clauss
Khadijah Harry
```

You could alternatively create an object and map

dictionaries to output rows. You'll need to use the DictWriter class whose details are shown below:

```
class csv.DictWriter(f, fieldnames, restval='',
extrasaction='raise', dialect='excel', *args,
**kwds)
```

Like DictReader(), DictWriter() has seven parameters, but the first two are mandatory. Below is an example on how to use DictWriter() to write to a file in Python:

```
import csv
with open('high_scores.csv', 'w') as file:
    fieldnames = ['player A', 'player B']
    writer = csv.DictWriter(file, fieldnames)
    writer.writeheader()
    writer.writerow({'player A': 102920, 'player B': 120121 })
    writer.writerow({'player A': 119822, 'player B': 130921 })
    writer.writerow({'player A': 190219, 'player B': 150292 })
    writer.writerow({'player A': 192911, 'player B': 167292 })
    writer.writerow({'player A': 199919, 'player B': 178281 })
```

Here's how the high_scores.csv file looks when opened:

player A	player B
102920	120121
119822	130921
190219	150292
192911	167292
199919	178281

Here's what's happening with the program. The csv module is imported, the open() function is called to open the file, and the 'w' option is used to write to it – a list with the name filenames is created and passed as the second argument of the writer() function.

The csv.DictWriter class is used and file and filenames are passed as the arguments. Once done, everything is setup using the writeheader() function which writes a row with the field names. Then, several rows of data are written using the writerow() function. Alright, now that we've learned the basics of the csv module let's do something more interesting with the data then just merely reading and outputting it. Let's refer back to the data sheet in software_company.csv: http://bit.ly/20IY9Ba

What if we wanted to transform the data so that instead of being in the following format:

John Q	lead software engineer	80000
Anon Pikes	software engineer	75000
Marky M	business analyst	65000
Don L	software tester	65000
Lisa Z	web developer	72000
Melissa J	marketer	50000
Daniel K	human relations	55000

It's now in this one?

```
John Q
lead software engineer
80000
----------
Anon Pikes
software engineer
75000
----------
Marky M
business analyst
65000
```

```
----------
Don L
software tester
65000
----------
Lisa Z
web developer
72000
----------
Melissa J
marketer
50000
----------
Daniel K
human relations
55000
```

Well, before we get to this there's still a couple of things we need to learn such as accessing the values in software_company.csv.

By analyzing the file we can see that we have tabular data that's represented in a row-by-column format. There are seven rows and three columns to be exact. Also, we know that a list is a common data structure in Python that uses subscript notation to access elements. Therefore, we can read in the file and then convert it to a list as shown in the following code snippet:

```
import csv
with open('software_company.csv', 'r') as file:
    data = csv.reader(file)
    employees = list(data)
```

As you can see the csv file is opened in read mode, and then the csv.reader() function is called with file passed as the argument. Then, the data variable which now stores the file is converted to a list by being passed into the list() mutable sequence type. Now that we've converted the file object into a list we can now access various portions of the csv file by using subscript notation. For example, let's say that you want to access the third row of the file, or the following:

```
Marky M        business analyst        65000
```

What would be the correct code to use for this? The answer is as follows:

```
employees[2]
```

What about if you want to access this column?

```
lead software engineer
software engineer
business analyst
software tester
web developer
marketer
human relations
```

This is a little more trickier as lists don't have columns therefore there's no built in function for this. Therefore, let's observe the pattern of the software_company.csv file. What we see is that the elements we want is in fact the second element of each row. Therefore, one solution is to extract these specific elements by looping through the csv file and then creating a new list that consists of only the second elements of each row. Here's how the code looks:

```
import csv
with open('software_company.csv', 'r') as file:
```

```
data = csv.reader(file)
data = list(data)
row, col = 0,1
while row < len(data):
    print(data[row][col])
    row+=1
```

Here's what the output looks like when printed:

```
lead software engineer
software engineer
business analyst
software tester
web developer
marketer
human relations
```

So by observing the data in the csv file we can see that they're seven rows and three columns. Therefore, translated into Python code we want to access the row starting at 0, and then iterate to the upper bound which in this case is 7. We want to update the code so that we print the name of the employee followed by their title, salary, and then a dashed line to separate the data. Here's the pseudo code of what we're going to do:

```
iterate through every row
print element in first column
print element in second column
print element in third column
print a dashed line
repeat above until end of the file
```

This logic translated into Python code is listed below:

```
import csv
```

```python
with open('/home/dougie/Desktop/software_company.csv',
'r') as file:
    data = csv.reader(file)
    data = list(data)
    row, col = 0,0
    while row < len(data):
        col = 0
        while col < 3:
            print(data[row][col])
            col+=1
        row+=1
        if row == 7 and col == 3:
            break
        else:
            print("------------")
```

JSON module

JavaScript Object Notation aka *JSON* is a light-weight format used for data interchange. It's important to know that JSON is language agnostic and many programming languages can generate and parse JSON data. JSON was created in the early 2000s by American programmer Douglas Crockford who also made many contributions to JavaScript such as *JSLint* and *JSMint*. The official media type for JSON is application/json, and the filename extension is .json. In the olden days web services used XML as the primary format for transmitting data, but since the arrival of JSON there's a new sheriff in town. JSON is used primarily to transfer data between web app and server and is an alternative to XML. The issue with XML is that it can be verbose at times adding to bandwidth consumption and download times. If you're using AJAX to make data requests then you can easily send and retrieve objects as JSON strings. You can learn more about JSON from the official website json.org.

JSON syntax

Even though you may think you need to know JavaScript in order to use JSON, you don't. JSON uses conventions that's familiar to programmers accustomed to the C-style family of languages. For example, there are two built in data structures in JSON which are:

- **name/value pairs**: In many programming languages this is known as an object, record, struct, dictionary, hash table, keyed list, or associative array.
- **ordered list of values**: In many programming languages this is known as array, vector, list, or sequence.

These universal data structures which can be found in a myriad of programming languages take the form of object, array, value, string, and number in JSON. Below is an explanation of each one.

Object

An object is an unordered pair of name/value pairs. The object starts with a left curly brace ({) and ends with a right curly one (}). Each name is followed by a colon, and each name/value pairs are isolated by a comma. Below is an example of an object in JSON.

```
{
  "name": "Clark Kent",
  "alias": "Super Man",
}
```

Array

An array is an ordered collection of values. An array begins with a left square bracket ([) and ends with a right square one (]). Values are separated by commas as shown in the example below:

```
["superhero", "journalist"]
```

Value

A value can be a string (double quotes), number, object, array, true, false, or null. A sample of some of these types in a JSON object is shown below:

```
{
  "daily_planet_salary": 65000,
  "balance": "23000",
  "withdrawal": -1500.98,
  "job_status": null,
  "still_super_hero": true
}
```

String

A string in JSON is similar to a string in C or Java. It's a sequence of 0 or more Unicode characters enclosed in double quotes. Backslashes are permitted and characters are represented as a single character string. Below is an example of a string in JSON:

```
{
   "main_villian": "Lex Luthor",
   "best_friend": "James Olsen",
   "cousin": "Kara Zor-El"
}
```

Number

A number can be any digit, therefore integers or floating point numerals are permitted. Below is an example of JSON code that consists of numbers:

```
{
 "number_of_movies": 8,
 "number_of_comics": 1600,
 "times_died": 15
}
```

Below is an example which puts everything together.

```
{
 "name": "Clark Kent",
 "alias": "Super Man",
 "nationality": [
    "United States",
    "Krypton"
 ],

 "occupation": [
```

```
      "journalist",
      "superhero"
  ],

  "main_villian": "Lex Luthor",
  "best_friend": "James Olsen",
  "cousin": "Kara Zor-El",
  "daily_planet_salary": 65000,
  "balance": "23000",
  "withdrawal": -1500.98,
  "job_status": null,
  "still_super_hero": true,
  "wife": "Lois Lane"
}
```

You can use a tool like *JSON Lint* to validate your code:
https://jsonlint.com

Now that we understand the basics of JSON we can learn more about the JSON module in Python. The core methods that you should be concerned with are dumps(), dump(), loads(), and load().

dumps → returns a string representing a JSON object.
loads → returns an object from a string that maps to a JSON object.
load and dump → read/write from file instead of a string.

The dumps() method takes an object and produce a string as shown below:

```
>>> import json
>>> car = {'mileage': 159238.3}
>>> json.dumps(car)
```

```
'{"mileage": 159238.3}'
```

The `loads()` method is used when you want to convert a string into a JSON object. For example, let's assume that you have the following JSON snippet below but in the form of a string in Python:

```
days = """
{
  "mon": "Monday",
  "tues": "Tuesday",
  "wed": "Wednesday",
  "thurs": "Thursday",
  "fri": "Friday"
}
"""
```

The string can be decoded via the `loads()` method.

```
d = json.loads(days)
d["mon"]
d["tues"]
```

```
for x in d.values():
    print(x)
```

The `load()` method is used when you want to read a file. It will take a file like object, read data from it, and use that string to create an object.

```
with open('superman.json') as s:
    a = json.load(s)
>>> a
```

The contents of `superman.json` will be printed to the console:

{'name': 'Clark Kent', 'alias': 'Super Man', 'nationality': ['United States', 'Krypton'], 'occupation': ['journalist', 'superhero'], 'main_villian': 'Lex Luthor', 'best_friend': 'James Olsen', 'cousin': 'Kara Zor-El', 'daily_planet_salary': 65000, 'balance': '23000', 'withdrawal': -1500.98, 'job_status': None, 'still_super_hero': True, 'wife': 'Lois Lane'}

BeautifulSoup is What's for Supper

BeautifulSoup is a Python package that's useful for parsing HTML and XML documents. It does this by creating a parse tree for parsed pages that can be used for extracting data from HTML. It was created by LA-based software engineer Leonard Richardson. BeautifulSoup is a quite sophisticated parser as it can successfully parse broken HTML. You can view the source code repository of BeautifulSoup: http://bit.ly/2zWFkRn

At the time of publication BeatuifulSoup4 is the recommended version of the package as its predecessor, BeautifulSoup3 is no longer being developed. Also, for brevity purposes we'll rename BeautifulSoup4 as BS4.

Highlights of BS4

- BS4 is a 3rd party library provided by Crummy.com and therefore is not included in the Python Core. That means that we may have to install it.
- BS4 automatically converts incoming documents into Unicode and outgoing documents to UTF-8.

- According to the website BS4 parses anything you give it.
- It's commonly used with urllib or the requests package to extract precise elements from a web page, aka web parsing.
- It's compatible with third party parsers like lxml and html5lib.

Installing BS4

Below are the various ways in which you can install BS4 on your machine. Installation on Debian or Ubuntu:

```
$ apt-get install python-bs4 (for Python 2)
$ apt-get install python3-bs4 (for Python 3)
```

Installation through pip

If you have pip installed on your machine which is a package manager then you can install BS4 through it by using the following code:

```
$ pip install beautifulsoup4
```

Also, there's a Python module known as easy_install which is a bundled with a whole bunch of setup tools and can assist with downloading, building, installing, and managing Python packages. To install BS4 with easy_install use the following:

```
$ easy_install beautifulsoup4
```

If you don't have pip or easy_install on your system then you can try a different approach. You can download the BS4 source tarball and install it with setup.py. You can download the source here: http://bit.ly/2Pn5CD0

Once the code is downloaded you can run it using the following syntax:

```
$ python setup.py install
```

If for some odd reason all of the above fail then there's still one more remedy. The BS4 license permits you to package the entire library with your application. Therefore, you can download the tarball file into your app's code base and therefore use BS4 without having to install it. BS4 comes with support for the HTML parser which is included in the Python's standard library, but it also includes support for other third party parsers.

```
$ pip install beautifulsoup4
```

Also, there's a Python module known as easy_install which is a bundled with a whole bunch of setup tools and can assist with downloading, building, installing, and managing Python packages. To install BS4 with easy_install use the following:

```
$ easy_install beautifulsoup4
```

If you don't have pip or easy_install on your system then you can try a different approach. You can download the BS4 source tarball and install it with setup.py. You can download the source here: http://bit.ly/2Pn5CD0

Once the code is downloaded you can run it using the following syntax:

```
$ python setup.py install
```

If for some odd reason all of the above fail then there's still one more remedy. The BS4 license permits you to package the entire library with your application. Therefore, you can download the tarball file into your app's code base and therefore use BS4 without having to install it. BS4 comes with support for the HTML parser which is included in the

Python's standard library, but it also includes support for other third party parsers. One of the popular choices is the lxml parser which you can download through the terminal, easy_install, or pip as follows:

```
$ apt-get install python-lxml
$ easy_install lxml
$ pip install lxml
```

Another option is to use the html5lib parser which is a pure Python parser that parses HTML in a similar fashion to a web browser. The below examples show how to install the html5lib parser:

```
$ apt-get install python-html5lib
$ easy install html5lib
$ pip install html5lib
```

To confirm that everything has been installed properly type the following into the terminal:

```
from bs4 import BeautifulSoup
```

If no error propagates then that means that you're ok. Once confirmed, it's time to move on to more interesting matters.

Stirring up the Soup - Analysis of the BS4 API

To parse a document with BS4 you can pass it to the BS4 constructor - you may pass in a string or a file. We'll be using the following HTML code as an example of what we'll be parsing. It's just a simple HTML code that includes a prospective list of the college football teams for the current season.

```html
<!doctype html>
<html>
<head>
 <meta charset="utf-8">
 <title> NCAA Football Rankings </title>
 <meta name="Simple website that displays top 20 NCAA rankings.">
 <meta name="author"content="Dougie Doug">
 <link rel="stylesheet" href="css/styles.css?v=1.0">
</head>
<body>

<h1> NCAA Football Rankings </h1>

<p> This is a ranking of the top 25 NCAA football rankings. </p>

<table border="1">
<tr>
    <td>Rank </td>
    <td>Team</td>
    <td>Rec</td>
    <td>Pts </td>
    <td>Trend </td>
</tr>
<tr>
    <td>1 </td>
    <td>Alabama (61) </td>
    <td>0-0 </td>
    <td>1621 </td>
    <td>----    </td>
</tr>
<tr>
    <td>2 </td>
```

```
        <td>Clemson (3) </td>
        <td>0-0 </td>
        <td>1547 </td>
        <td>----    </td>
    </tr>
    <tr>
        <td>3 </td>
        <td>Ohio State (1) </td>
        <td>0-0 </td>
        <td>1458 </td>
        <td>----    </td>
    </tr>
    <tr>
        <td>4 </td>
     <td>Georgia </td>
        <td>0-0 </td>
        <td>1452 </td>
        <td>----    </td>
    </tr>
    <tr>
        <td>5 </td>
        <td>Oklahoma </td>
        <td>0-0 </td>
        <td>1288 </td>
        <td>----    </td>
    </tr>
    <tr>
        <td>6 </td>
        <td>Washington </td>
        <td>0-0 </td>
        <td>1245 </td>
        <td>----    </td>
    </tr>
    <tr>
        <td>7 </td>
        <td>Wisconsin </td>
        <td>0-0 </td>
        <td>1243 </td>
        <td>----    </td>
    </tr>
    <tr>
        <td>8 </td>
        <td>Miami </td>
        <td>0-0 </td>
        <td>1091 </td>
```

```
   <td>----    </td>
</tr>
<tr>
   <td>9 </td>
   <td>Penn State </td>
   <td>0-0 </td>
   <td>1050 </td>
   <td>----   </td>
<tr>
   <td>10 </td>
   <td>Auburn </td>
 <td>0-0 </td>
   <td>1004 </td>
   <td>----    </td>
</tr>
</table>
 <script src="js/scripts.js"></script>
</body>
</html>
```

You can view the contents of site.html on GitHub here:
http://bit.ly/2Nx3AOM

Below is an example of how to pass in the contents of site.html into the BS4 constructor:

```
from bs4 import BeautifulSoup
with open('site.html') as f:
   soup = BeautifulSoup(f)
>>> soup
```

The BeautifulSoup library is imported and then the HTML file is opened via the with() function. The BeautifulSoup() constructor is created and the file, or f is passed to it. Next, the file is printed which displays the contents of the HTML tag. BeautifulSoup transforms the HTML document into a complex tree of Python objects. Most Python developers will typically only have to worry about four objects which are: Tag, NavigableString, BeautifulSoup, and Comment.

Tag:

This corresponds to a XML or HTML tag in the original document. An example of it in action is listed below:

```
with open('site.html') as f:
   soup = BeautifulSoup(f)
   tag = soup.tr
>>> tag
<tr>
<td>Rank </td>
<td>Team</td>
<td>Rec</td>
<td>Pts </td>
<td>Trend </td>
</tr>
```

The tag object generates the content between the tag. Tags have many attributes and methods with the main ones being its name and attributes. You can change the tag's name as shown below:

```
tag.name = 'row'
>>> tag.name
'row'
>>> type(tag)
<class 'bs4.element.Tag'>
```

Name:

Every tag has an associated name which can be accessed via .name.

```
>>> soup.tr.name
'tr'
>>> soup.td.name
'td'
>>> soup.body.name
'body'
```

Attributes:

A tag may have an infinite number of attributes. The tag can be accessed directly using the .attrs attribute, or you can access specific elements of it using subscript notation. Both methods are shown below:

```
with open('site.html') as f:
    soup = BeautifulSoup(f)
    tag = soup.link
```

```
>>> tag
<link href="css/styles.css?v=1.0" rel="stylesheet"/>
```

```
>>> tag.name
'link'
```

```
>>> tag.attrs
{'rel': ['stylesheet'], 'href': 'css/styles.css?v=1.0'}
```

```
>>> tag['rel']
['stylesheet']
```

```
>>> tag['href']
```

```
'css/styles.css?v=1.0'
```

Also, you can add, remove, and modify a tag's attributes by treating the tag as a dictionary as shown in the following example:

```
>>> tag['stuff'] = 'New attribute'
>>> tag['test'] = 'Test'
>>> tag.attrs
{'rel': ['stylesheet'], 'href': 'css/styles.css?v=1.0',
'stuff': 'New attribute', 'test': 'Test'}
>>> del tag['test']
>>> tag.attrs
{'rel': ['stylesheet'], 'href': 'css/styles.css?v=1.0',
'stuff': 'New attribute'}
```

Next, you can get more granular with attributes which is a little bit more tricker. In HTML 4 certain attributes can have more than one value but HTML5 removes some of them, and define a couple more. One of the most common multi-valued attribute is class; a tag can have more than one CSS class. Other examples are rel, rev, accept-charset, headers, and accesskey. BS4 represents the values of a multi-value attribute as a list.

NavigableString:

A NavigableString is the bit of text between tags. For example:

```
with open('site.html') as f:
    soup = BeautifulSoup(f)
    tag = soup.p
    tag.string
```

' This is a ranking of the top 25 NCAA football rankings. '

You can replace a string with another by using the replace_with() method as shown below:

```
>>> tag.string.replace_with("Top 25-Rankings")
' This is a ranking of the top 25 NCAA football rankings.'
>>> tag
<p>Top 25-Rankings</p>
```

How to explore the tree

Now that we've covered some of the aspects of BS4 it's time to implement more practical examples. Again, the following code snippet shows how to import a file and convert it to BS4:

```
from bs4 import BeautifulSoup
with open('site.html') as f:
    soup = BeautifulSoup(f, 'html.parser')
```

Once everything is setup you can run various methods on it.

Going down

The easiest way to navigate across a parse tree is to use the tag names. For example, to navigate via the head, meta, title, link, h1, and table tags respectively type the following:

```
>>> soup.head
>>> soup.meta
>>> soup.title
>>> soup.h1
>>> soup.table
```

The problem with using this method is that you'll only access the first occurrence of the tag. What if you wanted to access all of the occurrences of the tag? You can do it by using the find_all() method. For example, using just soup.td would yield:

```
>>> soup.td
<td>Rank </td>
```

However, using soup.find_all('td') would yield:

```
>>> soup.find_all('td')
[<td>Rank </td>, <td>Team</td>, <td>Rec</td>, <td>Pts </
td>, <td>Trend </td>, <td>1 </td>, <td>Alabama (61)
</td>, <td>0-0 </td>, <td>1621 </td>, <td>---- </td>,
<td>2 </td>, <td>Clemson (3) </td>, <td>0-0 </td>,
<td>1547 </td>, <td>---- </td>, <td>3 </td>, <td>Ohio
State (1) </td>, <td>0-0 </td>, <td>1458 </td>, <td>----
</td>, <td>4 </td>, <td>Georgia </td>, <td>0-0 </td>,
<td>1452 </td>, <td>---- </td>, <td>5 </td>,
<td>Oklahoma </td>, <td>0-0 </td>, <td>1288 </td>,
<td>---- </td>, <td>6 </td>, <td>Washington </td>,
<td>0-0 </td>, <td>1245 </td>, <td>---- </td>, <td>7
</td>, <td>Wisconsin </td>, <td>0-0 </td>, <td>1243
</td>, <td>---- </td>, <td>8 </td>, <td>Miami </td>,
<td>0-0 </td>, <td>1091 </td>, <td>---- </td>, <td>9
</td>, <td>Penn State </td>, <td>0-0 </td>, <td>1050
</td>, <td>---- </td>, <td>10 </td>, <td>Auburn </td>,
<td>0-0 </td>, <td>1004 </td>, <td>---- </td>]
```

Big difference! A tag can have many attributes, and this aggregate of attributes are called it's *children.* You should be able to access the tag's children by using .contents as shown below:

```
>>> soup.tr.contents
['\n', <td>Rank </td>, '\n', <td>Team</td>, '\n',
<td>Rec</td>, '\n', <td>Pts </td>, '\n', <td>Trend
</td>, '\n']
>>> soup.head.contents
['\n', <meta charset="utf-8"/>, '\n', <title> NCAA
Football Rankings </title>, '\n', <meta name="Simple
website that displays top 20 NCAA rankings."/>, '\n',
<meta content="Dougie Doug" name="author"/>, '\n', <link
href="css/styles.css?v=1.0" rel="stylesheet"/>, '\n']
```

You can access specific elements of .contents by sub scripting it as shown below.

```
>>> soup.head.contents[0]
'\n'
>>> soup.head.contents[1]
<meta charset="utf-8"/>
>>> soup.head.contents[2]
'\n'
>>> soup.head.contents[3]
<title> NCAA Football Rankings </title>
>>> soup.head.contents[4]
'\n'
>>> soup.head.contents[5]
<meta name="Simple website that displays top 20 NCAA
rankings."/>
```

Also, the normal operations of a list can be applied to contents. Below is an example on how iterate over the children of a tag. You can use the children generator to do this.

```python
from bs4 import BeautifulSoup
with open('site.html') as site:
    soup = BeautifulSoup(site, 'html.parser')
    head_tag = soup.head
    for child in head_tag.children:
        print(child)
```

Going up the tree

Every tag and string has a parent, aka the tag that contains it. You can discover an attribute's parent by using .parent. For example, in site.html the parent of the meta tag is the head tag since the meta tag is enclosed within the head. An example of how to access the parent of a tag is listed below:

```python
>>> meta_tag = soup.meta
>>> meta_tag.parent
```

Also, just in case you've been wondering what's the parent of a top-level tag like <html> then it's the BeautifulSoup object itself as displayed by the following code snippet:

```python
>>> html_tag = soup.html
>>> type(html_tag.parent)
<class 'bs4.BeautifulSoup'>
```

Here's more proof. Since the BS4 object is at the very top of the directory, it has no parent. Therefore, the following will print None:

```
>>> print(soup.parent)
None
```

You can iterate over all of the elements of a parent with
`.parents` as shown below:

```
para = soup.p
for parent in para.parents:
    if parent is None:
        print(parent)
    else:
        print(parent.name)
```

```
body
html
[document]
```

Sideways

You can parse a BS4 object going sideways. For example,
let's look at the following code snippet:

```
from bs4 import BeautifulSoup
with open('site.html') as site:
  soup = BeautifulSoup(site, 'html.parser')
>>> soup.prettify()
```

The last statement calls the `prettify()` method on the
soup object. When an object is so-called *pretty*, this causes
the object's siblings to appear at the same indexation level.
You can therefore use the `.next_sibling` and
`.previous_sibling` attributes to navigate between page
elements that's on the same level as the parse tree, hence

why this is considered parsing from *side-to-side*. For
example, look at the following:

```
from bs4 import BeautifulSoup
with open('site.html') as site:
    sib_soup = BeautifulSoup(site, 'html.parser')
    sib_soup.td.next_sibling.next_sibling
```

```
...
<td>Team</td>
```

Parsing XML with BS4

Parsing a XML document is pretty much the same as parsing
a HTML document with the exception that you pass in a
second parameter which will be the XML parser: lxml or
html5lib. Let's assume that the XML document that we want
to parse is listed below. Its filename is book.xml.

```
<?xml version="1.0" encoding="UTF-8"?>
<book>
 <title> Java for Newbies. </title>
 <subject> computer programming </subject>
 <author> Dougie Doug </author>
 <description> Master the basics of the Java core
</description>
 <pages> 292 </pages>
 <price> 180 </price>
</book>
```

Below is an example on how to read in the XML file via BS4
using the xlib parser:

```
from bs4 import BeautifulSoup
with open('book.xml') as site:
```

```
  soup = BeautifulSoup(site, 'lxml')
>>> soup
```

To read in the XML file using the html5lib parser is pretty much the same except you use html5lib as the second parameter indicated below:

```
soup = BeautifulSoup(site, 'html5lib')
```

Once the file is read into memory and passed into the BeautifulSoup constructor the file be manipulated. The way to manipulating XML is similar to manipulating the HTML file. For example, look at the following coding snippets:

```
>>> soup.title
<title> Java for Newbies. </title>
>>> soup.subject
<subject> computer programming </subject>
>>> soup.author
<author> Dougie Doug </author>
>>> soup.description
<description> Master the basics of the Java core </
description>
>>> soup.pages
<pages> 292 </pages>
```

In BS4, there's no class for parsing XML. Instead, you pass in XML as the second argument in the BeautifulSoup constructor.

Chapter V Optimized

In this chapter we learned some cool ways on how to manipulate and parse data in Python. Since CSV files are so common these days it made sense to learn about the CSV module in Python and how to use it to read/write, and transform data. Another file format that's very ubiquitous these days is JSON so we learned how to use the JSON module as well. Then, we learned about the BeautifulSoup4 library which is a very popular choice when it comes to parsing HTML and XML files in Python.

Chapter V Coding Challenge

Let's parse some data. Using the same HTML file we worked with in this chapter (site.html) lets use BS4 to traverse all of the elements in it. Here's how the output of the first several tags should look:

```
html
head
meta
title
meta
meta
link
body
h1
p
table
tr
td
td
td
td
td
```

Chapter V coding solution: http://bit.ly/2RCIBxA

Chapter VI: The Tour de Python Library

Erik Drost - Cleveland Public Library — Image - CC BY 2.0

When you download Python a standard library is shipped with your distribution. This library is in essence a bevy of built in modules that provides access to system functionality such as Graphical User Interface (GUI), Input/output (I/O) operations, and statistics. There's a lot of cool things that you can do with the Python library and every serious Python programmer should have a general understanding of it. This chapter will provide a high-level overview of some of the core modules of the Python library.

Quick Tkinter Tutorial

Tkinter is the defacto package for GUI programming in Python. It's easy to configure since it's bundled in many of the binary distributions of Python. To test that Tkinter is already installed on your machine use the following statement:

```
python3 -m tkinter
```

If a simple Tkinter GUI pops on the screen then you're in good shape. If not, then you'll have some installing to do. To install Tkinter via Debian/Ubuntu use the following command:

```
sudo apt-get install python3-tk
```

The following code snippet shows how to create a simple window using Tkinter:

```
import tkinter as tk                                          1
class Window(tk.Frame):                                       2
def __init__(self, master=None):                              3
tk.Frame.__init__(self, master)                               4
        self.grid()                                           5
        self.create()                                         6
    def create(self):                                         7
        self.quitButton = tk.Button(self, text='Close',
                            command=self.quit)                8
        self.quitButton.grid()                                9
```

```
app = Window()                                        10
app.master.title('first tkinter program')            11
app.mainloop()                                        12
```

Below is the output:

Close

Figure: 6.0. Simple Window in Tkinter.

Below is an explanation about what's happening:

1) Imports the tkinter package as tk.
2) Creates a class that extends Tkinter's Frame class.
3) Creates a constructor for the Window class.
4) Calls the Frame constructor.
5) Places widgets within a two dimensional grid. The grid() method is one of the three built in layout managers: pack() and place() are the other two.
6) Calls the create() method. See steps #7-9...
7) Defines the create() method.
8) Creates a button with the text *Close*; the command self.quit means that the button exits when clicked on.
9) Sets the layout for the button.
10) Creates instance of the Window class.
11) Sets the title of the Frame.
12) Calls the mainloop() method to execute the GUI.

This program takes an object oriented approach to building the Window. However, as you know Python also allows imperative style programming so this style is not mandatory. Here's the above program translated using the imperative paradigm:

```
from tkinter import Frame, Button
root = Frame()
root.grid
```

```
b = Button(text='Close', command=quit)
b.grid()


root.mainloop()
```

One is not necessary better than the other, it's more a matter of preference and who's going to see your code. If you plan on building a big application that source code will be modified by many programmers then the OOP approach makes more sense due to better maintainability. However, if it's a one-off small script then the procedural style will be just fine. Here's another program that shows how to position buttons in a Frame using Tkinter:

```
from tkinter import Tk, RIGHT, RAISED, BOTH
from tkinter.ttk import Frame, Style, Button
class Window(Frame):
    def __init__(self):
        super().__init__()
        self.create()
    def update_text(self):
        print("Hello")
    def create(self):
        self.master.title("Buttons")
        self.style = Style()
        self.style.theme_use("alt")
        frame = Frame(self, relief=RAISED, borderwidth=2)
        frame.pack(fill=BOTH, expand=True)
        self.pack(fill=BOTH, expand=True)
        close = Button(self, text="Close",
command=self.quit)
        close.pack(side=RIGHT, padx=5, pady=5)
        message = Button(self, text="message",
command=self.update_text)
        message.pack(side=RIGHT)
```

```
root = Tk()
root.geometry("250x250")
app = Window()
root.mainloop()
```

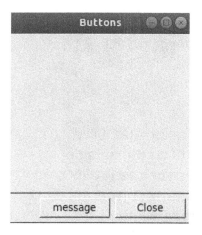

Figure 6.1: Buttons Tkinter.

Let's concentrate on the create() function as here is where most of the new concepts are at. You can set the theme for a widget with the theme_use() method. This portion of the code is responsible for setting the theme of the frame:

```
self.style = Style()
self.style.theme_use("alt")
```

First, an instance of Style is created, and them from there the style.theme() method is called. It's important to know that ttk is a theming layer for tkinter. It contains a collection of themes that can be applied to widgets. To figure out what themes you have available you can use the following code snippet:

```
>>> import tkinter.ttk as ttk
>>> style = ttk.Style()
>>> style.theme_names()
```

('clam', 'alt', 'default', 'classic')

The classic option is default. You can test out other themes so that you can gain a better sense of which look you prefer. Let's take a look at the following snippets of code:

```
frame = Frame(self, relief=RAISED, borderwidth=2)
frame.pack(fill=BOTH, expand=True)
self.pack(fill=BOTH, expand=True)
```

In the above code a Frame is created with relief=RAISED – the relief style refers to simulated 3D effects on the outside of a widget. There are other relief attributes such as RAISED, SUNKEN, and GROOVE, but some of them may not work depending on the OS you're using – the borderwidth attribute sets the the border which in this case is 2. The frame.pack() method places a number of widgets into a column. The fill and expand attributes deals with spacing. The fill attribute determines what to do with the empty space provided, and the expand attribute determines what to do with extra space. Therefore, if there's two frames with expand=True, then tkinter will give half of the extra space to one space, and the other half to the remaining one. These statements create the buttons:

```
close = Button(self, text="Close", command =
self.quit)
close.pack(side=RIGHT, padx=5, pady=5)
message = Button(self, text="message", command=
self.update_text)
message.pack(side=RIGHT)
```

The button with the text "Close" exits when clicked on. The button with the text "message" has the following: command = self.update_text

This points to a method that simply prints *Hello* to the console when clicked on. The dimensions of the window is set with this instruction:

```
root.geometry("250x250")
```

This allows you to shape the size of a window in width and length format.

File handling in Python

The ability to process files is important in computer programming and Python has many specialized functionality for this. Here's a sample code snippet of file handling in Python:

```
>>> file = open("file.txt", "w")
>>> file.write("This is the first message\n")
26
>>> file.write("This is the second message\n")
27
>>> file.close()
>>> file = open("file.txt", "r")
>>> one = file.readline()
>>> two = file.readline()
>>> file.close()
>>> print(one + two)
This is the first message
This is the second message
```

The open() function creates a text file named file.txt in

the current working directory, and since it's opened with the "w" mode that means it's ready to *write*. Two messages are written to the file and then the same file is opened and its content is read into memory using the "r" mode. The contents of file.txt are then printed.

Regular expressions

A **regular expression** is a a sequence of characters that represents a search pattern. Regexes are kind of a language within itself, but mastering them are worth it due to their widespread use. Python uses the regular expression syntax that's similar to Perl. Below is how you can create your first regex in Python in three easy steps:

1) Import the regular expression module using the following statement: import re
2) Create the expression for the regex
3) Use a built in function in the re module to look for a match

Here's a quick regex example:

```
>>> import re
>>> regex = re.compile("hello")
>>> test = "hello world this is a simple regex
example. I'll put another hello for good measures."
>>> regex.search(test)
<_sre.SRE_Match object; span=(0, 5), match='hello'>
```

As you can see the output returns an object that provides various information about the regex such as where it appears in the string and also the specific match. If no match is found then the function would return None. In the above example it's probably not necessary to compile the pattern into a regular expression object. However, when the expression will be used several times throughout the program then compiling is the way to go. Most searches for regexes are not as simple as looking for one word; it may need to look for multiple variations of a word. There's something known as *special characters* in regexes that affect how regular expressions around them are interpreted. Let's look at some code to gain a better idea:

```
>>> import re
>>> regex = re.compile("dog")
>>> print(regex.search("Doggy Dog"))
None
```

The reason for this is because the regex is looking only for the lowercase version of the word "dog." How do we fix this? Should we make two separate regexes, one to match the lowercase version, and another for the uppercase derivative? That's feasible but not optimal. A better solution is to make use of Python's built in classes to modify the regexes to be more inclusive of the search terms that we want. In other words the regexes needs to be modified to cast a wider net as shown below:

```
>>> import re
>>> regex = re.compile("(D|d)oggy (D|d)og")
>>> find = regex.search("Doggy dog")
>>> find
<_sre.SRE_Match object; span=(0, 9), match='Doggy dog'>
```

The parentheses means that you're grouping terms together. The other symbol is the pipe (|) and translates to OR. Therefore, if you want to search for x or y terms then you can use the pipe symbol. To make better sense of this let's dissect the above regular expression one-by-one. For example, look at:

(D|d)oggy (D|d)og

The (D|d) portion means to look for the pattern D followed by *oggy* or d followed by *oggy*. Therefore, *Doggy* or *doggy* are both valid entries. The second part states to look for D followed by *og*, or d followed by *og*. Therefore, valid inputs are *Dog* or *dog*. Look at the following code fragment:

```
regex = re.compile("^the")
>>> print(regex.search("the movie rocks!"))
<_sre.SRE_Match object; span=(0, 3), match='the'>
```

The caret special character (^) matches the start of a string. What do you think the following will print?

```
>>> print(regex.search("this is the best movie
ever"))
```

The answer is None because *the* is not at the beginning of the string. Another way that you could create an exact match for the is to use both the carat and the dollar symbol ($). The $ symbol matches the end of the string or just before the newline. Therefore, you could update the regular expression as follows:

```
>>> regex = re.compile("^the$")
```

Look at how the following expressions evaluate:

```
>>> regex.search("the")
<_sre.SRE_Match object; span=(0, 3), match='the'>
>>> print(regex.search("the best movie ever"))
```

None

If you want to define a set of characters and look for the match of any of them in a set then you can use the square brackets []. An example of it in action is listed below:

```
>> import re
>>> regex = re.compile("[a-zA-Z0-9|;|<|>]")
>>> regex.search("??????????<>ajsjso8822")
<_sre.SRE_Match object; span=(10, 11), match='<'>
```

The syntax a-z, A-Z, and 0-9 mean any lower case letter, any uppercase letter, or any digit – so it is essence denotes any alphanumeric character. The pipe symbol (|) represent or, so this regex translated into English states: *any lower case, or upper case letter, or any digit, or semicolon, or a less than sign, or a greater than sign.* If you want to specify how many copies a regex should match you can use curly braces such as {m}. Look at the following snippet:

```
>>> regex = re.compile("[a-z]{3}")
>>> regex.search("??????????<>a(j_s^j#ssdsso8822")
<_sre.SRE_Match object; span=(20, 23), match='ssd'>
```

Here's what this new regular expression states translated into English: *"A match shall be made if there's precisely three occurrences of any letter in a row."*

That's why the first occurrence of a letter is not a valid match, because it has to be three letters in a row, not just one. There's other ways in which you can control the number of repetitions for a regex. You can use + to cause the regex to match 1 or more repetitions of the proceeding regex, * to cause a match for 0 or more repetitions, and {m,n} to cause the regex to match m through n repetitions of a regex. Here's a quick pop quiz on regexs. Which of the following are valid input for the following regexs?

1) RE = abc+
a) abccc
b) abc
c) abbbc
d) abcabc

2) RE = abc{3}
a) abc
b) abcabc
c) abc*3
d) abcabcabccc

3) RE = [a-z]{3}
a) 111
b) a9aa83jmnsm02
c) aaa
d) b292k0202

Answers:
1 all
2 c and d
3 b and c

There's a lot of nifty things you can do with regular expressions and it's a good idea to get good at them because at some point in your software engineering career you'll probably need to use them. I would recommend checking out good ole Python docs for more details on how to use them: http://bit.ly/2pIIfIZ

Mathematica

If you need to do some mathematical operations then use the **math module**. If you're familiar with C then you should have no issue getting acquainted with the library asap as the math module provides access to the underlying C library functions for floating point math.

```
import math
def algebra_formula(a,b):
    return math.pow(a,2) + 2*a*b + math.pow(b,2)
>>> algebra_formula(5,2)
49.0
```

The math.pow(a,2) statement provides the same result as a**2. Here's a formula that combines trigonometry with algebraic functions.

```
>>> import math
>>> math.sqrt(5*math.pi + math.sin(10)*
math.cos(math.pi))/(math.sqrt(5))

1.8028857079048763
```

The math module provides a plethora of mathematical goodies such as power and logarithmic functions, trigonometric functions, special functions like gamma, and constants like *e*.

The *random* module implements pseudo-random number generators for various distributions. The central function in this module is random() which generates a random float uniformly in the semi-open range [0.0,1.0].

```
>>> import random
>>> random.random()
0.8389043939711015
```

```
>>> food = ["pizza", "burgers", "sandwich", "cake"]
>>> random.choice(food)
'burgers'
>>> items = ['coffee', 'hot chocolate', 'tea']
>>> random.shuffle(items)
>>> items
['hot chocolate','tea', 'coffee']
# returns floating point between numbers
>>> random.triangular(1,10)
8.471049332878868
['tea', 'coffee', 'hot chocolate']
>>> random.sample(range(100),10)
[71, 49, 17, 82, 10, 57, 62, 0, 22, 24]
>>> random.betavariate(5,5)
0.22311876948268683
```

Statistics module Python

This module provides various functions for calculating mathematical statistics. Here's a list of some of the functions:

mean()	Arithmetic mean, or average
harmonic_mean()	Harmonic mean
median()	Middle value
median_low()	Low median of data
median_high()	High median of data
median_grouped()	50th percentile of grouped data
mode()	Most common value of discrete data

Figure 6.2:Statistics functions.

```
import statistics
>>> scores = [50.6, 98.87, 65, 81, 85, 92, 65.2, 81,
79.4, 87, 93]
>>> statistics.mean(scores)
79.82454545454546
>>> statistics.harmonic_mean(scores)
77.0490469155837
>>> statistics.median(scores)
81
>>> statistics.median_low(scores)
81
>>> statistics.median_high(scores)
81
>>> statistics.median_grouped(scores)
81.25
>>> statistics.mode(scores)
81


import statistics, random
start = 10
end = 101
nums = []

while start < end:
    i = random.randrange(start)
    nums.append(i)
    start+=1

>>> nums
[2, 5, 8, 9, 9, 8, 10, 2, 6, 5, 7, 16, 15, 15, 11, 19, 14, 14,
3, 7, 10, 5, 27, 17, 15, 9, 28, 32, 31, 34, 30, 23, 2, 28, 42,
33, 7, 20, 36, 5, 41, 2, 13, 3, 26, 31, 25, 26, 35, 21, 49,
16, 48, 8, 15, 57, 4, 14, 26, 16, 27, 24, 43, 34, 73, 40, 6,
40, 4, 63, 51, 36, 56, 81, 76, 80, 3, 29, 24, 38, 36, 60, 14,
78, 56, 10, 35, 44, 91, 7, 48]
```

```
>>> statistics.mean(nums)
26.395604395604394
```

Accessing the web

There are various modules that you can use to access the web and process internet protocols. Two of the more simpler ones are `urllib.request` for retrieving data from URLs, and smtplib for sending emails. If you need to use a higher-level HTTP client interface then it's recommended to use the requests package. Below is a brief tutorial that covers the basics of `urllib.requests` and smtplib. The central function of the `urllib.request` module is the `urlopen()` function which is similar to the open() function with the difference that the `urlopen()` function can only open URLs for reading, and no seek operations are available. This is not the module to use if security is critical as the server certificate is not validated.

```
urllib.request.urlopen(url[, data][, timeout])
```

The only mandatory argument is url aka the website, while the other two are optional. This module supports several functions such as read(), readline(), readlines(), fileno(), close(), info(), getcode(), and geturl(). Below is a simple illustration of how to use the `urllib.request.urlopen()` function.

```
from urllib.request import urlopen
with
urlopen('https://en.wikipedia.org/wiki/Python_(prog
ramming_language)') as webpage:
  print(webpage.read())
...
```

Here's some of the code snippet that'll get printed:

```
b'<!DOCTYPE html>\n<html class="client-nojs" lang="en"
dir="ltr">\n<head>\n<meta charset="UTF-8"/>\
n<title>Python (programming language) -
Wikipedia</title>\
n<script>document.documentElement.className =
document.documentElement.className.replace( /(^|\\
s)client-nojs(\\s|$)/, "$1client-js$2" );</script>\
n<script>
...... etc ......
```

The above code snippet creates a webpage object for the Wikipedia url, reads in the content, and then prints it. It prints all of the content including the HTML and text. The smtplib module defines an SMTP client session object that can be used to email to any Internet machine with an SMTP or ESMTP listener daemon. The following script shows how to automatically send an email using gmail. This could be useful if you're creating an app and it needs to fire an email after a certain action has occurred. An example of how to send an email using smtplib with a gmail account is listed below:

```
import smtplib

server = smtplib.SMTP('smtp.gmail.com', 587)
server.starttls()
server.login("email", "password")

message = "Put your message here!"
server.sendmail("email", "to address", message)
server.quit()
```

This example uses Gmail and depending on your account security you may need to update your settings to grant access for *less secure apps*. This uses the **Simple Mail Transfer Protocol** (SMTP) which is an internet standard for transmitting email. The number 587 is a port that's reserved for mail servers. In order for the email to be sent you must authenticate your email by entering the correct information in server.login() – therefore, put in your email address and password in the arguments. Last, put in all of the information such as the email, to address, and message in the server.sendmail() function.

Dates and Time

The datetime module supplies classes for manipulating dates and times. To create an object that represents a date in the point of time look at the following code snippet:

```
>>> import datetime
>>> d1 = datetime.date(1900,1,31)
>>> d2 = datetime.date(1990,5,20)
>>> print(d2-d1)
32981 days, 0:00:00
```

The default format for the datetime.date class is year, month, and day. You can change the order in which the date is printed by using Python's strftime directives.

```
>>> d1.strftime("%m/%d/%y")
'01/31/00'
```

Data Compressing

Archiving and compressing is supported by modules such as zlib, gzip, bz2, lzma, zipfile, and tarfile. This is useful for applications in which data compression is needed. Lets look at the zlib module:

```
>>> import zlib
>>> content = b"this will get zipped up!"
>>> zipped = zlib.compress(content)
>>> zlib.decompress(zipped)
b'this will get zipped up!'
```

Performance Management

Python provides a measurement tool so that you can determine the relative performance of different approaches to the same problem. The module to use is timeit which provides a simple way to test the performance of bits of Python code:

```
>>> python3 -m timeit "10*10"
100000000 loops, best of 3: 0.0138 usec per loop
>>>python -m timeit "10**100"
100000000 loops, best of 3: 0.0155 usec per loop
python3 -m timeit "'-'.join(str(x) for x in range(500))"
10000 loops, best of 3: 140 usec per loop
```

The -m stands for microseconds which are one million in a second. Also, by default *100000000* loops are ran.

Quality Control

As you probably figured out already a ton of things can go wrong when it comes to programming so that's why it's important to have some quality insurance mechanisms built into your code. Luckily, Python provides several features that can assists programmers with bulletproofing their code. The doctest module is a tool for scanning a module and validating tests embedded in a program's docstrings. This makes the documentation more useful as it provides the user with an example of how to use the function, and allows the doctest module to ensure that the code is coherent with the documentation.

```python
def largestnumber(*args):
    """computes the largest number in a list:

        >>>
print(largestnumber(0,1,29,9,3,6,7,2,283,2,8,65,56))
        283

    """
    nums = [ ]
    i = largest = 0
    for x in args:
        nums.append(x)
    while i < len(nums):
        if nums[i] > largest:
            largest = nums[i]
            i += 1
        else:
            pass
            i += 1
    return largest


import doctest
doctest.testmod()  # automatically validate the embedded
```

tests

```
print(largestnumber(0,1,29,9,3,6,7,2,283,2,8,65,56))
```

There's also the unittest module that enables a more comprehensive set of tests to be maintained in a separate file. The unittest testing framework was inspired by JUnit and is similar to the testing frameworks in other languages. It comes jam packed with a myriad of features such as test automation, sharing of setup and shutdown code for tests, and aggregation of tests into collections. An example of it in action is shown below:

```
import unittest
import LargestNumber as l

class TestFunction(unittest.TestCase):
    def test_largest_number(self):
        self.assertEqual(l.largestnumber(1,2,3,4,5),5)
        self.assertEqual(l.largestnumber(39,2,29,12,0,993,
73,56,233),993)
        self.assertEqual(l.largestnumber(20,54,87,100,65,8
7,98), 100)
        self.assertIs(l.largestnumber(1,2,3,4,5,6),l.large
stnumber(1,1,1,3,6))

if __name__ == '__main__':
    unittest.main()
```

The LargestNumber module is imported which contains the largestnumber() function named LargestNumber. It then uses the assertEqual() and assertIs() functions of the class to test the functions.

Operating System Interface

This module provides a myriad of operating system dependent functionality. You can read or write to a file by using the open() function, or if you need a way to manipulate paths then you can use the os.path module. There's a lot of cool things that you can do with this:

```
>>> import os
>>> files = os.listdir('/home/doug/Desktop/Files')
>>> files
['MOCK_DATA.csv', 'DATA.csv', 'emails.txt',
'Songs', 'Text', 'Vowels.txt',
'worldcitiespop.txt', 'nycstats.xml',
'customers.csv', 'Characters.txt', 'letters.txt',
'en.xml']

for x, y in enumerate(files):
  print(x,y)
...
0 MOCK_DATA.csv
1 DATA.csv
2 emails.txt
3 Songs
4 Text
5 Vowels.txt
6 worldcitiespop.txt
7 nycstats.xml
8 customers.csv
9 Characters.txt
10 letters.txt
11 en.xml

# gets current directory
>>> os.getcwd()
'/home/doug/Desktop/Files'
```

The sys module provides access to some variables that's

used or maintained by the interpreter.

```
>>> import os
>>> sys.float_info
sys.float_info(max=1.7976931348623157e+308,
max_exp=1024, max_10_exp=308,
min=2.2250738585072014e-308, min_exp=-1021,
min_10_exp=-307, dig=15, mant_dig=53,
epsilon=2.220446049250313e-16, radix=2, rounds=1)
>>> sys.executable
'/usr/bin/python3'
>>> sys.platform
'linux'
>>> sys.api_version
1013
>>> sys.maxsize
9223372036854775807
```

There are many ways in which you can read in user input. However, if you want to read into the standard input use stdin as shown in the small code snippet below:

```
import sys
for line in sys.stdin:
    print(line.rstrip())
```

Multi threading in Python

A **process** is a program that has been loaded into memory and running. Early computers only allowed one program to be executed at a time. Modern computers are a whole new ballgame. Multiple programs can be loaded into memory and executed concurrently. A system is therefore a collection of processes; operating system processes executing system code and user processes executing user code. You can see all of the processes that's running on a windows machine by opening up the terminal and typing the following:

```
ps  -A
```

On Linux, type:

```
ps
```

A program by itself is not a process, it must be loaded into memory before it's labeled as a process. The two common ways in which a program is loaded into memory is by double clicking the executable file, or entering the filename of the executable file on the command line such as program.exe, or code.out.

The advent of a concept known as *multi threading* allows modern computers to take advantage of technology advances and speed computers up. For example, most modern computers provide facilities for a process to contain multiple threads of control. A **thread** is a basic unit of CPU utilization, and it contains things like a thread ID, program counter, register set, and a stack. It can communicate with other threads that are in the same process and share resources such as the code and data sections. If a process has multiple threads of control then it can perform more than one task at a time. Multi threaded programs provides many benefits such as allowing the program to be more responsive, using less resources, and increasing scalability – a multi threaded computer that has multiple CPUs (cores) can facilitate *parallelism*. Below is a sample program that illustrates the concept of threading in Python:

```python
import threading
import time
def loop():
  for i in range(1,20):
      time.sleep(2)
      print(i)


>>> threading.Thread(loop()).start()
```

The above code snippet prints the numbers 1-19 but with the caveat that there's a two second pause between each print. The time.sleep() function is responsible for this.

Database programming in Python

Python includes support for a myriad of database programming languages such as MySQL, PostgreSQL, SQLite, and Oracle. The Python standard for database interface is Python DB-API (PEP 249). The database interface that we're going to discuss is sqlite3. What's cool about this is that it's included in Python right out of the box, and doesn't require any additional setup. SQLite can be used for internal data storage, or the code can be ported to larger database systems such as Oracle. This is useful if you want to use the database on disk, and don't need a full-fledged solution. The below code snippet shows how to get started with sqlite3 in Python.

```python
import sqlite3

conn = sqlite3.connect(":memory:")
cursor = conn.cursor()
cursor.execute('CREATE TABLE products (id INTEGER
PRIMARY KEY, title TEXT)')
```

```
cursor.execute('INSERT INTO products VALUES(?,?)', (1,
"Java For Newbies"))
cursor.execute('INSERT INTO products VALUES(?,?)', (2,
"Python Programming"))
cursor.execute('INSERT INTO products VALUES(?,?)', (3,
"JavaScript Programming"))
cursor.execute('INSERT INTO products VALUES(?,?)', (4,
"Master 12 Programming Languages"))

cursor.execute('SELECT * FROM products WHERE id =?',
(1,))
row = cursor.fetchone()
id = row[0]
title = row[1]
print("id =", id, "name", title)
conn.commit()
```

The sqlite3 module is imported and then a connection to the database is established. The parameter is ":memory:" which means that it's an in memory database. You could alternatively create a database locally by passing a database name in the parameter. For example, the following statement could be used instead:

```
conn = sqlite3.connect("store.db")
```

If you create a physical database on disk then the issue is when you rerun the program errors could manifest if the database already exists. When it's created in memory, the database is re-created each time. Once the database is created, the connect() function is used to make a connect object – this is used to execute statements on the database. A table is created and then four rows are inserted into the table. It's important to know that the symbols ?,? are used as placeholders which is considered *Pythonic* as it helps makes the database secure adding an additional layer against SQL injection. The fetchone() function is used to retrieve the rows which can be accessed using subscript notation.

Chapter VI Optimized

In this chapter you got a quick rundown of some of the cool features in Python available in the standard library. As you have probably grown to learn over the course of this book, Python is a freaking powerful language and the standard library is a testament to that. Once you have a strong grasp of the syntax and semantics of Python, mastering the Standard Library is the next logical step in your quest to becoming a Python Guru. You can check out the epic contents of the Python library here:
https://docs.python.org/3/library

Knowing every single item is most likely not necessary (even though it would be nice and makes you a stronger candidate for job interviews). I would recommend checking out the standard library to see any modules that interests you and just dig-in and start building something cool with it.

Chapter VI Coding Challenges

Use the sympy module to solve the following system of equations. Sympy is a module in Python for symbolic mathematics. You can learn more about sympy here:
https://www.sympy.org/en/index.html

Coding challenge 1: Solve the following system of equations.

$3x + 15y = 5$
$10x - 3y = 15$
$4y + 10z = 19$

Coding challenge 2: Solve: $x^2 - 16$

Coding challenge 3: Solve: $x^2 + 10x - 5$

Challenge 4: Graph the following equations using SymPy:

a) $x^2 + 5$
b) $x^3 + 10x + 5$
c) $a^7 - 5a$

Chapter VI coding solutions: http://bit.ly/2y87oS2

INDEX

Thanks for Reading! Reviews are Welcomed :)

Your opinion does matter! Make sure to leave a review for *Become A Python Developer* and let your voice be heard. Leaving reviews are fun, gives your bragging rights on social media, and helps you master the content of the book you're reading. Go ahead and review *Become a Python Developer* now and let your voice be heard.

www.ingramcontent.com/pod-product-compliance
Lightning Source LLC
Chambersburg PA
CBHW082117070326
40690CB00049B/3603